be kept

ENGINEERS AND IVORY TOWERS

⊓⌐⊓⌐⊓⌐⊓⌐⊓⌐⊓⌐⊓⌐⊓⌐⊓

HARDY CROSS

Engineers and Ivory Towers

EDITED AND ARRANGED BY

Robert C. Goodpasture

FIRST EDITION

McGRAW-HILL BOOK COMPANY, INC.

NEW YORK · TORONTO · LONDON

1952

ENGINEERS AND IVORY TOWERS

Library of Congress Catalog Card Number: 51-12937

FOREWORD

His office is long and dimly lit, walled in on two sides by an expanse of books and reports reaching to the ceiling. A full width of windows at the far end of the room is partially blocked by overhanging ivy, but admits enough light to silhouette the figure at the desk. Hundreds of men who have worked within reaching distance of Hardy Cross can recall the familiar sight of him gazing thoughtfully through the window, thick cigarette smoke drifting slowly over his head.

What is the nature of this teacher of engineering on whose library shelf *The Theory of Elasticity* stands side by side with *Abe Lincoln in Illinois, The Holy Bible* and *Alice in Wonderland?* Why have men traveled from the Americas, Europe and Asia to sit in his classroom? How have his opinions on engineering and education come to be so widely respected?

Hardy Cross was class valedictorian when graduated, at the age of seventeen, from a small college that had a long tradition of liberal culture: Greek and Latin, the English classics, mathematics and science, philosophy and religious history. After brief excursions into literature and chemistry, he studied civil engineering at two larger schools, at one as an undergraduate and at the other as a graduate student.

His intimate association with engineering education continued to develop over the ensuing years as he held professorships at three universities, the present one being Strathcona Professor of Civil Engineering at Yale. Thus he has enjoyed firsthand familiarity with six different schools of higher learning. Because of his personal contact with graduate students over many years, he has had an unusual opportunity to meet and know men from numerous other educational institutions. All the while, this background has been tempered with years of experience in the planning, design and building of engineering works.

The engineering world has long acclaimed Hardy Cross for his professional achievements and bestowed upon him honors from professional and academic organizations. A citation by the American Society for Engineering Education indicates that he received its Lamme Medal in 1944 ". . . for his development of revolutionary methods of analysis in structural engineering; for his application of these methods to the rigorous training of civil engineers; for his insistence on the great responsibilities of the individual teacher and his scorn of the superficial in education; . . ."

The philosophy of this book has been developed by the author over the past thirty years and has appeared in both published and unpublished form. The editor, while a student at Yale, voluntarily undertook to organize an extensive collection of these works of Professor Cross.

Source material was of both a technical and non-technical nature, available in various magazine articles, society papers, transcribed speeches, longhand classroom notes and graduate lectures. It was originally written for many different groups under widely divergent conditions, and organizing, editing and rearranging were necessary. This book contains about one-sixth of the material initially accumulated. The wording of many papers was necessarily modified somewhat in the interest of uniformity of style. In certain instances, however, the reader will note that original stylings have been intentionally preserved because to change them would materially affect the meaning.

Many persons have evidenced interest and offered more than mere encouragement as work on this book progressed. Its publication is full realization of the value of their help. Dr. William S. Livingston, Assistant Professor in the Department of Government at the University of Texas, deserves particular recognition for his timely suggestions concerning the manuscript, as does Professor Frederic T. Mavis, of the Carnegie Institute of Technology, whose assistance has helped to make possible the publication of this book.

Robert C. Goodpasture

New York City
January, 1952

Firm Foundations for Towers

ENGINEERING, SCIENCE AND THE HUMANITIES

"Engineering is the art of planning for the use of land and air, and for the use and control of water; and of designing, building and operating the works and machines needed to carry out the plan."

DEFINITIONS are a fetish with some, but defining terms does not always lead to definiteness of ideas. Engineering is the art that deals with the application of materials and material forces. The use of science is a means to that end. The purpose of engineering is service to mankind.

Pure science deals with problems involving fewer variables than does engineering and often involves a narrower range of variation than is found in engineering. To say that a man thoroughly trained in theoretical physics and chemistry is thereby properly trained to be a good engineer is highly misleading. Science as such should have nothing to do either with use or convenience. Science tries to find out the facts about materials and actions. There is considerable authority to support the opinion that great scientists do not follow quite the order of procedure in arriving at discoveries that they follow later in proving that their discoveries are true. This merely means that in creative science there is very distinctly an element of art, just as in art there is usually some science, or at least some system. Eventually in the most highly developed creative minds the two merge, but in the conventional literature and in ordinary affairs the two can be more or less distinguished. The systematized, formalized procedure called science, which is supposed to lead inevitably to unquestionable results, contrasts with the flexible independent creative instinct which produces art. A further distinction is that science seeks truth and should test itself only against truth. Art is concerned with the attainment of itself whether that end be beauty or usefulness. It uses all available means to attain its ends.

Art is creative, full of life, and can adapt itself to new ideas. Science tends to become more fixed in its methods, in its norms of thought, in its method of statement; with elaborated terminology it tends to develop a methodology. But this is the popular concept of science rather than that developed in the minds of the great creative scientists.

It has always been important that people understand clearly the nature, the types of problems and the processes used by engineers. They use any fact or theory of science, wherever and however developed, that contributes to their art. If a knowledge of physics, of chemistry, of meteorology, of mathematics is useful in attaining the ends in view, engineers will go to endless trouble to master these sciences for their purpose.

"One test is worth a dozen expert opinions"; on the other hand, someone else has said that "no test is worthy of credence unless supported by an adequate theory." Engineers can, unless they adopt a narrow and distorted view of learning, see and weigh the truth of these conflicting views.

Engineers are not, however, primarily scientists. If they must be classified, they may be considered more humanists than scientists. Those who devote their life to engineering are likely to find themselves in contact with almost every phase of human activity. Not only must they make important decisions about the mere mechanical outline of structures and machines, but they are also confronted with the problems of human reactions to environment and are constantly involved in problems of law, economics and sociology. It is fortunate that the engineer does not usually bother to clutter up these problems of human relations with technical, academic designations.

Engineers are guided by the facts of scientists, but their answers are not controlled by the physical facts alone. They are trying to use the facts, to manage them, if you will, to assemble them into new relations. There cannot be a more misleading view than that which pictures engineers as driving inevitably by mathematics or laboratory process to an unique solution of their problems; their solutions are rarely unique. Engineering is not mathematics, although it makes use of many mathematical processes. Engineers almost everywhere and all the time have one identifying trait; they want to put down some figures, to make a chart, to draw a plan. Engineers put down many figures, but they put them down as a guide for their thinking, not as an answer to their problem. They want evidence; they want scale on the problem; they want some plan as to where they are going and what will probably happen when they get there. The work of the engineer is by nature synthetic, although it has often ceased to be treated as such and this must again come to be recognized. It consists of putting together fragments from human relations, from science, from art, from craftsmanship to produce new assemblages. Simply making an analysis of all the elements, all the data of the problem, does not mean a solution has been obtained. These data must be put together, made into a new assembly that involves a large imaginative element, put together with due respect to the relative importance of the elements and to the probability of simultaneous occurrence; and all this must be done with some intuitive vision of what is wanted and of what can be got. Then, and then only, has there been a solution of an engineering problem.

There are always many ways of building, several ways of overcoming the obstacles. Some are best from the point of view of economy of materials, others from economy of men or time. Some are better because the result is more useful and some are better because the result meets more nearly the demands of convenience. There is often justification for building some transportation system such as a subway, not because people must have it, but because people want it. Engineers need not especially ask whether people should have it. If the demand is there, it is for the engineer to solve the problem and also to appraise the sacrifices involved.

This picture of engineering is not the one with which most

laymen are familiar. They believe that engineering work is done in a perfectly mechanical way, that engineering is a result of the inflexible application of formulas to physical phenomena; they have an impression that in this field scientific laws are very clearly known without exceptions. These laws, they think, are embodied in charts, tables and equations that represent facts about which there is no question and from which conclusions follow with unfailing accuracy. Those who have closely examined engineering thought know that most curves are lined with question marks and that the formulas are often merely a basis for discussion. Non-scientists think science is infallible, especially if stated in mathematical symbols. They do not know that the scientific laws that are of universal application are quite frequently true because the terms are defined in such a way as to make them true.

The laymen now extrapolate this concept of science and engineering. They have read that this is an age of science, that human welfare has been immensely promoted by science; their fancy runs to automobiles, airplanes, radio, television. The material world is being transformed and transformed rapidly. But the transformation must not be attributed to pure science alone. An essential element, perhaps the most important element, is the correlating faculty of the engineer rather than the pure research faculty of the scientist; such developments involve a large element of judgment, much uncertainty, much cautious trial and error.

Science standing alone contributes nothing to the welfare of mankind or to his illfare.

The glory of the adaptation of science to human needs is that of engineering. Misconceptions of this distinction between engineering and science are actually doing harm. In several cases the engineers are trying to do the work of the scientists because the scientists have failed to do it, and the scientists have failed to do it in many cases because they did not realize that the engineers wanted it done. There is a great need for very careful investigations by physicists on the action of materials under stress. No question can be raised of the great work done by engineers who are engaged in research in the properties of materials, but some of their problems should be referred, if possible, to the laboratories of trained physicists. The engineers should be relieved of the problem, or some parts of the problem, in order that they may devote their creative minds to other matters.

Laymen, observing that scientific or engineering methods—and they often fail to distinguish them—have altered appreciably the welfare of humanity, have now set out to improve humanity itself by a similar process. The procedure may be somewhat as follows: They collect statistical data showing the number of crimes per unit of population in various parts of a city and the distribution of taxable value of property per unit of population. Next a chart is plotted having as abscissas the taxable value, and as ordinates the criminal record. This gives a curve for which someone may even write an equation. They are then prepared to work with this equation, perhaps to differentiate with regard to taxable property and find out the minimum or maximum criminality per unit of taxable value.

This is a cartoon, but the point is this. Laymen feel that, having drawn this curve, they have a curve quite comparable, for example, to the endurance limit curve for steel and that the use and study of this curve promises quite as definite and tangible results as do data from engineering laboratories. The engineering mind is likely to be very skeptical of these data relating crime to poverty. Engineers recognize at once that the increase in crime may not be an effect of the poverty but that both may be concurrent effects of some other variable so that forcibly eliminating the poverty may not affect the criminality. Or the data on the incidence of crime may be undependable because of the methods of determining the amount of crime. Engineers are always critical of statistical data and regularly ask whether the indications of the data were not inherent in the method of collection.

The literature on fatigue of metals is both voluminous and bewildering. Results are influenced by the composition, treatment and past history of the metal. This is, of course, true of the laboratory specimen. When an attempt is made to apply even the more definite of these results to the design of a railway bridge, engineers encounter arguments that have continued for fifty years or more. How amazing, then, to find dogmatic statements about fatigue in human beings.

Some try to explain how in the future the methods of science are to be applied to the study and adjustment of human relations. In such thinking there may be three important errors for the too hopeful student. First, he misconceives the nature of science by ignoring the relative simplicity of the problems with which the pure scientist deals as compared with the complexity existing in the assembly of such problems by nature. Second, he confuses science and engineering and attributes the accomplishments of engineering, which are to a marked extent a result of inventive and synthetic power, to the accomplishments of science. Third, he errs in the concept of what this process of thinking is and how it accomplishes its results in the field of engineering science. He thinks that engineers arrive at truths by plotting charts, whereas engineers plot their charts to be considered as evidence in estimating probabilities. It is no wonder then that these methods of charts and formulas and mathematical symbols are being so often misused for selfish ends in a world obsessed by misconceptions of their use.

There are groups of self-styled engineers who are telling the country how valuable they are and how accurate are their conclusions. Take almost any general term, use it as an adjective and prefix it to engineer—social engineer, transport engineer, economic engineer, human engineer. These men attempt, often consciously though sometimes unconsciously, to give the impression that they deal with measurable data from which definite laws useful to mankind may be deduced. They often call this leadership. Real engineers are tired of these leaders, of men who scorn the details. Engineers usually know what they are trying to do.

Dr. Irving Langmuir, as President of the American Association for the Advancement of Science, presented a paper on this subject. Here a great scientist and engineer devoted a scientific address largely to pointing out the existing dangers in the overextension of what some conceive to be the scientific method. Particular reference was made to the misinterpretation of scientific procedures and the misinterpretation of evidence based upon procedures inapplicable in the field where they are used. The criticism was pointed apparently at sociologists and economists. The whole paper is impressive; especially so is the remark that there is a tendency to underrate the capacity of the human mind, and the strong plea for common sense in human affairs. At present one of the obsessions of many people is the antithesis that they conceive to exist between individualism and regimentation. The philosophic antithesis is rather old; consider the ecclesiastical arguments over predestination and free will. The engineer comes to understand as he grows up that there is here no necessary antithesis; that there can be much freedom with much regulation; that the regulation is bad if it destroys the originality; and that originality unchecked by evidence from the past and from common sense as to the present would best be checked by some regimentation.

Much has been written of the scientific method in engineering. The question is, is there a single scientific method in engineering or anywhere else? There are many methods of arriving at the truth, though often truth itself is uncertain because criteria are needed to determine what constitutes truth in special fields.

Engineering is essentially a craft. It is the glory of engineers that they are craftsmen, that they are artists, and while as good craftsmen they follow a systematic and orderly procedure, they are highly resistant and antagonistic toward overregimentation. They demand freedom of their art, freedom to recreate, to rearrange. Varying degrees of emphasis are given by different thinkers to the importance of human affairs, of genesis, of analysis, of synthesis—the creation of new concepts, the analysis of known phenomena, or the putting together of old things to make better things.

On the title page of the biography of that great leader in public health, William T. Sedgwick, is written: "He loved great things and thought little of himself. Desiring neither fame nor influence, he won the devotion of men and was a power in their lives; and seeking no disciples, he taught to many the qualities of the world and man's mind."

Standardization and Its Abuse

INTELLIGENT STANDARDS VERSUS
STANDARDIZED INTELLIGENCE

"The child has to be taught the words that correspond to things; the senior at college has lost the things that correspond to the words."

WHEN a structure is designed three quite obvious questions should be asked in succession: Do you want something? What do you want? How will you use it? These questions may not be asked or answered by one man, but all must be intelligently answered.

When something is wanted it is appropriate to ask why, when and where is it wanted, what sacrifice will be made to get it. The second question, "What do you want?" leads to the problems of what you have, of whether you can get what you want, and is it standard? The third question—use—involves problems of management, operation and finance.

"What do you have; what is available?" To face these questions we need a knowledge of types of construction, of materials available, of possible layouts, of general dimensions.

Consider the problem, "Can you get it?" suggested by the second question. This may be called "design" and is critical. It involves full study of construction procedures, of contractors, materials, labor, equipment, and time elements. Consideration must be given to appearance, architectural styles, harmony between style adopted and natural surroundings. Investigation should show the use and convenience of bridges and approaches, of buildings and industrial plants and yards and terminals. Economy, costs, values, and finally the structural elements in the problem must be reviewed in order to ensure strength, stability, stiffness and generally satisfactory performance of each structure without objectionable deterioration. All these factors contribute to the solution of the problem, "Can you get what you have decided that you want?"

Most literature in the structural field deals with strength and stability for the very good reason, not always obvious to the amateur, that if a structure is not sufficiently strong, it makes little difference what other attributes it has. One might almost say that its strength is essential and otherwise unimportant.

Various sources aid the engineer in determining strength. No one of them is more important than another. Analyses, tests, experience and such intuitive common sense as may be personally developed about structural stability; these are all helpful, but they can also be dangerously misleading. Evidence from the four sources rarely agrees completely. Great engineers are those who can weigh this evidence and arrive at a reasonable answer through judgment as to its dependability.

The materials to be used must be of standard manufacture; the advantages of standardization here should be obvious to all. Design loads, methods of analysis, allowable stresses; all must conform approximately to some standards which for certain types of work are narrowly circumscribed and for other types of work leave considerable latitude to the designer. There is a good deal of convenience in standardizing construction methods and materials as well as methods of fabrication and criteria for stability.

But there is another purpose of standardization here and in most engineering fields. It is helpful to think about engineering by distinguishing its creative and its routine features. It is clear that in all ages there have been men who planned physical developments; it makes little difference by what name they were called. These men were creative artists—those who built Babylon, drained the Pontine marshes, bridged the Thames at London or the Mississippi at St. Louis, planned works on the Merrimac or the Brandywine. As the size and complexity of projects increased, the time came when there was more work to do than men to do it or time in which to think out problems. It became desirable and even necessary to do then in the intellectual field what had been done earlier in the field of manufacture: to set up a series of routine procedures for analysis and for design. This meant the development of a series of formulas and rules and standards which could be followed within limits by men trained in that vocation, by men who had applied that formula in that way over and over until they could satisfactorily duplicate their results. With these standardized formulas and specifications and methods it became possible to use a greater number of men and men with less training to produce engineering works. There appeared then what was in effect an intellectual assembly line. It had the advantage that these young men could follow the standards and arrive at the same result whether they lived in Boston or Los Angeles and whatever the condition of their health or temper at the time they made the computations. In other words, work could be checked.

To that extent then something that was originally intelligent—the collecting and weighing of evidence and the thinking out of the criteria of stability and stiffness—had been standardized as on an assembly line. On this assembly line men could do over and over a specific operation in a clearly defined way.

Without these assembly lines and the use of mechanical brains it would be impossible to turn out the volume of work that comes from engineering offices today. At the same time most engineers are thoroughly familiar with the tragic results of this standardization when used without discrimination or control. They are conscious of this and have set up many safeguards against it.

The important point here is that some types of planning, designing and experimenting can be put on an assembly line and some types can be put on an assembly line of skilled brains only, but much of the most important work cannot be done by using fixed rules, standardized formulas or rigid methods.

Consider an example from a field commonly thought of as rather technical and standardized, the design of arches. Almost everyone has some interest in these if only because he has seen rainbows. The choice of layout of the arch is open to judgment. It should be beautiful, easy to construct, properly located. After these considerations have been settled a decision must be made concerning loads; no one can prophesy with certainty the loads that may come on a structure during its life. A digression into the loads and imposed deformations leads far afield—the development of vehicles of transportation, wind forces, temperature changes.

Allowable working stresses must be chosen. Again there is much uncertainty. Volumes of laboratory data have been accumulated, but the profession is still changing working stresses in concrete and steel.

Many men in many places in many ways are studying materials, how to mix concrete, how steel fails. References on fatigue and flow of metals pile up and, as so often happens, terminology often outruns reality. Speculations about the nature of failure and the phenomena that precede it continue. But there must be a bridge, an arched bridge; by the way, are we sure we want an arch at all?

Assume that all these matters have been settled; it has taken judgment, intelligence and art to settle them well. Now to dimension the structure. Engineering texts suggest that this is a very formal matter, that the procedure is to guess at dimensions, write some mathematical equations for given conditions of loading and find the stresses that result. If it is then found that the arch rib is overstressed, it should be changed; but this approach will not tell how to change it. One solution would be to make the crown deeper or shallower, but whether it should be changed depends on how much of the stress results from the weight of the rib, how much from that of the deck, how much from moving loads and how much from such things as changes of temperature.

After all these matters have been discussed, the analysis must be interpreted. Excessive dead-load stresses are not relieved by the plastic properties of the material, but excessive temperature stresses are much relieved by plastic flow; stresses from moving loads may be relieved by plasticity much or little.

It may be noted that in this field, commonly thought of as technically regimented where solutions are mathematical certainties, there is real need for imagination, vision and curiosity. Solutions may be far from unique. This situation is not peculiar to bridge design, but rather the example might as well have been chosen from any branch of engineering.

The assembly line can never replace the brain that has created it. Machines, methods and systems cannot be a substitute for men. Old techniques must be changed and often abandoned, new techniques developed. If entirely new techniques are to be developed, men must be trained ahead of time; the profession must tool up before the emergency, which means there must be a measure of standardization. Is that the function of the universities? There should be no dogmatic answer to that question. One thing is certain, however. There has always been, even in the worst of recessions, a shortage of men who could design the assembly lines or work well where assembly lines are ineffective; there has always been and will always be a shortage of creative thinkers in any field.

Medieval architecture was not standardized. That is one of its great charms. Dissymmetry is marked; apparently it is frequently intentional in the medieval cathedral. There is nothing very standard about Chartres or Mont-Saint-Michel. The little naked soul so prominent in sculptures of the Day of Judgment did not always outweigh the devil and his imps; in one of the column capitals at Saint-Lô the sculptor, perhaps suffering from morbid indigestion, reversed the procedure and thus caused great embarrassment for future curators.

In the field of structural design the effort to get intelligence through standardization has been carried pretty far. In reinforced concrete, for example, it has been necessary to set up elaborate standards. Out of this work came a narrowly circumscribed standardization of procedures, which is called "the theory of reinforced concrete" and to which unfortunate students are exposed. Few will question that the standardized theory of reinforced concrete is perhaps as complicated a bit of nonsense as has been conceived by the human mind. It does, however, work pretty well as a check on undiscriminating unintelligence.

In engineering there is no attempt to standardize unless there is some reason for it. Some, however, wish to standardize where there is no real advantage and so fasten for a long time upon the profession a complex assembly line that has characteristics of a cartoon. Standardization, as a check on fools and rascals or set up as an intellectual assembly line, has served well in the engineering world.

Unfortunately the objectives of standardization have often been misconceived outside the engineering world. Blind standardization on a huge scale may be tried under a cloak of humanitarianism and accompanied by the argument that thus engineering, which has become science, has revolutionized the physical world. In the end it will not work but in the meantime there may be much misery before redemption comes. When engineers standardize they at least confine their standardization to the pattern within which they wish to standardize—one thing for bridges, another for buildings, another for airplanes and another for streamlined trains.

It is practically impossible to put dates on engineering. It is equally hard to say that there are entirely new problems. The problems of today are in many respects the problems of hundreds of years ago, but these problems deal sometimes with new materials and always with different conditions. When a problem is all solved and the answer is very definitely known in the field of engineering, it is about time to investigate that problem again, because what is known is probably known for certain limited materials. But novelty should not be pursued for itself alone. The novelty often consists in merely doing another thing in about the same way that other things have been done before.

Unfortunately some glorify the pursuit of novelty for its own sake. Someone has analyzed stresses in a particular structural member by one arrangement of computations; another arrangement of the computations then constitutes an element of novelty. Unnecessary novelty in the field of art, as in the field of engineering, is something to be apologized for and not commended. Men must not be deceived into giving to dust that is a little gilt more praise than gilt o'er dusted. Amateurs clutter up the literature to produce the illusion of novelty where none exists and where none is wanted. This can be seen in art, philosophy, literature, economics and religion. The claim of novelty is used to cloak error and to spice insipidness.

While some men choose not to worship blindly at the shrine of novelty, it does not necessarily follow that they restrict their interests to the obvious. A clear and simple restatement of a fundamental principle may have profound influence. The virtue here is not due to any novelty of the rewording, but rather due to the simplicity and clarity of the contribution.

Engineering has, in most of its branches, been thinking out all of its problems again. This is not an indication that the laws of geometry or statics have changed or that there are any new principles about dynamics. However, new materials and new uses of old materials have been tried; new methods of using old principles have been invented. In nearly every field of engineering now there is a seething activity of invention, investigation and reinvestigation. Some of this is probably ill-directed. What is needed are men with ability to orient some of these investigations in a new way.

News, novelty, uniqueness is often dependent upon the fancy and conditions of the times. Long timber trusses are more news today than they were in 1850. Brunel used reinforced brickwork over a hundred years ago; the use of mechanical models is not by any means new; the principle involved in the deformeter gage comes from the last century; "soil mechanics" is a new name but the study of foundations, of soil pressures, of soil resistance is not a new thing. There was a period of cantilever construction, then a period of continuous construction, and later a reversion to the cantilever.

Extensive organized investigations in structures have usually resulted from some immediate problem, such as the large increase in the height of skyscrapers in the twenties, the Long Beach earthquake, the Miami hurricane, increased highway traffic, larger storage dams. Repairing the barn door does not imply building a new type of barn. A new development is often merely of temporary importance.

In general the objectives are flexibility of design and

simplicity of construction. Design should seek convenience or use or beauty of outline, and this design should result in simple and economical construction. Development of a solution may be due to an engineer's special knowledge of structural forms or to a construction man's ability to burn and weld. Sometimes a solution might be credited to the grace of the equipment manufacturer or perhaps to a field man who can mix better concrete.

The history of engineering, like that of structural development, represents the parallel growth of four elements: materials, methods used in field or shop, concepts used in design, and those pictures that make more definite and clear the elements in that design. Immediate necessity, often economic, dictates which of these elements develops and which lags in any decade.

Development and advancement are largely dependent upon research which, by necessity, deals with controlled study of small isolated details. There is usually a long period before such details can be assembled into generalizations. Many try to seize upon these details before they have been digested and apply them at once. What are supposed to be results of investigations are often incorporated in specifications and codes before the investigation itself has been completed, much less digested. There is, then, always the danger that immature conclusion will become "frozen" in practice and hence be reported as a "new development."

Yes, there is development and progress. In some fields

the development is slow. Men must learn to think more clearly in space and be less restricted to two-dimensional design. They must pay more attention to movements and vibrations. They need much more information on the properties of materials. Probably they need to reappraise seriously the importance of durability. A few need to be told that the pursuit of novelty does not always lead to progress.

The time has come in many fields to take stock. There is continuous production of analytical tools, continuous accumulation of data from tests, continuous construction of bigger and supposedly better machines and structures. But we need now to take stock of what we know, what we do not know, what we need to know and why. There must be more of this work in the future. It is difficult to do at all and very difficult to do well. The sympathetic interest of the research man and the scholar is needed. It must be done in the interest of education on the one hand and of practice on the other; it is wrong to continue indefinitely to add, add, add to the tools of knowledge, without combination or elimination.

Some Ivy and Some Ivory Towers

EDUCATION, TRAINING, SCHOOLING

"With blossomed furze, unprofitably gay."

DISTINCTION should be made between education, training and schooling; the distinction is not entirely pedantic. It is difficult to educate without training and equally hard to train without to some extent educating. But the two things are not the same. Everyone knows more or less what education is and everyone misinterprets it at times. Schooling is helpful in the process of education.

Many in America grew up in a tradition of overorganized, oversystematized methodology of knowledge. It often resulted in paralysis of initiative and sterility of imagination. By that philosophy every possible case must be formulated in advance. Consider a modification of Josh Billings's epigram: "It is better not to plan so much than to plan for so many things that never happen."

The purpose of education is to prepare a whole man to live a full life in a whole world. American colleges must produce men who can think out American problems in American ways. To do this they must turn out men who see America and American life as a whole and also see the relation of America to the world. The country cannot afford to depend on men who will bury themselves intentionally in some narrow aspect of that life.

This is a big order and never fully attained, but to say that a man is educated as an engineer or educated as a doctor, or as a lawyer, an educator, or an economist—that is to say that he is partly educated. These distinctions between the mental disciplines through which men grow into full life are frequently set up because of local limitations or for administrative purposes. Overemphasis on such distinctions is very bad.

In a way, education is a rather simple matter. Most men wish more information about their world and seek better correlation and interpretation of the information that they have. Good schooling may help much in guiding to information or illuminating correlation.

But schools are far from simple, and there lies the trouble. Libraries and laboratories, buildings and red tape, overlapping departments apparently closely related but really uncorrelated, elaborated administrative organizations, textbooks and techniques—these, in varying degree, characterize the schools. Much of this merely amuses the fancy of dilettantes without guiding to education.

There is little parallel in the real world for the rigid distinctions between departments of a university. They are the result of necessary organization that grows and grows into the overorganization that the graduate soon learns to recognize in corporation or professional society. Departmental differentiation thus reaches the state of the good lady who thanked God that, though her church had saved only two sinners during the year, the horrid old congregation down the street had not saved a single damned soul. Teachers sometimes seem more anxious to damn some other field of learning than to illumine the pathway of education.

It is easier to teach rules than it is to train judgment; therefore, when teachers get tired in the schools they are likely to revert to rules. These can be taught to students and it is possible to give examinations and grades on them. But it requires high art to teach and examine on judgment; let anyone who doubts this try to do it. Consequently college curricula, whether in structural design or literary criticism, tend to degenerate into compilations of rules, regulations, cases and classes unless these curricula are constantly revitalized. The same thing may be said of activities outside the schools.

But the rules must be taught as well as the judgment, and college is a good place to teach many of the rules. Ripe judgment comes only with experience. The thoughtful man concedes it is well for student, teacher, and practicing engineer frequently to ponder Tredgold's definition of engineering, "The art of directing the great sources of power in nature for the use and convenience of man." Those whose vaulting ambition for leadership would o'erleap the painful need of accurate information must be reminded that they cannot well direct that of which they know little—no, not even by the most hopeful art.

A university has a trinity of influence: through the faculty and its work; through the campus life of student societies and publications; and finally through something which should be deeper, older, more stable—the spirit and tradition that pervade the campus, the lecture hall, the laboratory.

This spirit that drives on to the pursuit of truth results from the accumulated greatness of a group of scholars who have learned to care very much whether things are done well or ill, to care very much whether work is useful or useless. And they have learned to judge truth without appeal either to popular vote or to intellectual dictatorship.

If the young men can "go places," let them go. Constant nursing and guiding in colleges is not the paramount need, but rather a great impersonal light leading men on. Unfortunately that light can fade in the garish klieg lights of too much ballyhoo, of too many popular conclusions, of too much sense that is too common.

Colleges often swing from periods of cerebral malnutrition through inspirational debauches to periods of intellectual indigestion. Success carries within itself the elements of failure—unless it's profoundly sound. Too many old men, set in their ways, are ready to guide. And to guide often means to rule, to suppress, to kill. And so young men are sent on petty errands with few new ideas.

Honest pursuit of truth is very well worth while for the sake of truth and for the sake of honesty. And consistent honesty in the pursuit of truth will produce plenty of individualism, the type of individualism that is not imitative or conventional, the type of individualism that is not captivated by the latest fad. A great university is a group of honest scholars. Such a group of honest scholars will produce honest students, honest thinkers and honest men. And such men will not be blown about by every whispering breeze of fancy.

Education must not become formalized, but the educators should clarify its objectives and maintain freedom in seeking those objectives. The progress of students is often unduly burdened with details of learning. Some engineers go so far as to say that the function of the technical schools is to teach a man to do a particular job in a particular way. No! The purpose of schools is not to meet the needs of particular industries, and in this one finds support from many leaders of industry. The function of the universities is to turn out intelligent men with some knowledge of practical fields rather than to turn out non-intelligent men with detailed knowledge of limited fields.

Many of the best educated men never saw the inside of a college until they went there in later life to give commencement addresses or to sit as members of the corporation. But today there is a growing obsession for academic credits and guinea stamps of learning and a growing confusion between literacy, training, learning and wisdom. Standardization in fields outside of engineering is apparently inherent in animal nature; habit and imitation are inherent in human make-up. But most people welcome a break from this standardization; many come eventually, if it goes too far, to hate it bitterly.

Education should give men an opportunity, with some content and purpose, to develop freely their intelligence, to think some things out themselves, to arrive at conclusions new at least to them. The textbooks do not help much here. Many texts are written in stilted terminology, contain too many overelaborated definitions and state so-called fundamental principles that do not exist.

One of the latest slogans is "education for citizenship." When, please, was education for anything else but citizenship—but does this mean standardized citizenship and is it to be your standard or the standard of some bureaucrat? Is the student to be indoctrinated with all sorts of nostrums, the knowledge of which is alleged to be prerequisite to good citizenship? William Graham Sumner's "Forgotten Man" was a nonconformist, an ordinary fellow attending to his business as well as he could but, outside of any technical requirements, forming his own decisions. But such men —these little fellows—become the butt of ridicule or focus of attack of enthusiasts who insist on standardization. Sumner's essay closes with the apposite remark that "the forgotten man—who is frequently a woman—works and possibly prays, but you may be sure that he always pays."

This is not a criticism of any particular form of education. Probably there is not one single proper form. It is a
mistake for a man to go to college for narrowly vocational
objectives, unless he clearly recognizes that what he is getting is training and not education. Education in structural
engineering is not necessarily more narrowing than classical
studies of "terminations in T in Terence." It may be accepted
that some bad education is worse than none and more bad
education is worse than less. This needs to be stated and
restated.

To send people to college with too vague purposes simply
to learn standardized forms of fragmentary knowledge is
dangerous. Hitler taught the world how very dangerous the
pretense of education may become, not dangerous in the
old-fashioned sense but dangerous in the ghastly horrible
sense of modern war and modern Europe. It rejects the concept of the free man thinking through the world in which he
lives as God gives him the intellect to do so. Many still have
an abiding faith that this dream of the unstandardized free
man persists, but he is easily imitated to an unsuspecting
student who seeks light where there is no light. In great
schools—not large, great—free men work in an atmosphere
of great thoughts, of great faiths and of great dreams; but
the thoughts need not be couched in stilted or artificial or
technical language, the faith need not be placed in the whims
of dictators, the dreams need not be nightmares. Liberal

education is still indefinite to many; there are too frequent
and varied definitions of it. Many who argue for it actually
get far from liberal education. The dream of a whole man
in a whole world must not be swallowed up in vocationalism,
overspecialization, pompous nomenclature. The purpose of
education must be service and not self-promotion. The
dream of an individual who stands squarely on his own
feet, whose intelligence is independent of dictum and
dogma, who looks with faith at the future and smiles, this is
the dream that carried us across the continent, and it must
not be lost. The flood tide of progress always comes slowly
from far back through creeks and inlets of individual
thought.

The Pharisee prayed, "O God! I thank thee that I am not
as other men are." Most of our universities are in keeping
with a great American tradition and recognize an obligation to guide and inspire national thinking. It should not be
forgotten that they are centers of general education, that
while specialized interests may furnish impetus to search
for knowledge, it is the whole man that should be educated
to live a full life. Some seek at college that which is not
there to be sought, like the man who looked for some lost
trinket where he knew he had not left it because the light
was better where he was looking. Some believe in baptism
because they have seen it done and forget the inner wisdom
implied in education while admiring outward visible man-
ifestations.

By the side of the Pharisee in the temple a publican prayed, "God be merciful to me, a sinner." God enlighten our ignorance, keep our thinking simple, keep our education straightforward. A college that keeps that faith will truly educate; one that forgets it fails. Through humility we may continue to "instruct youth in the Arts and Sciences who through the blessing of Almighty God may be fitted for Publick employment both in Church and Civil State." That concept of a university is still pretty sound.

The general characteristics of inflation are easily recognized. It tries to make the reality—the goods—seem more valuable by making more plentiful the thing—money—that is exchanged for the goods. There is then more money, and men feel richer because they get more money and have more money.

The fallacies of inflation are vigorously deplored by many educators who are themselves enthusiastic for inflation in education. In the educational world the goods is the training of the man, and the thing through which he acquires that training is equipment, personnel and curricula. Some educators appear to think that if the number of courses and classes available is increased, then the training will be better and more valuable.

Apparently, many are beginning to see that universities will gain rather than lose by adopting a less costly and pretentious scale of doing things. You may admire the man who made two blades of grass to grow where one grew before, but not in your flower beds. He who sets up two courses where one grew before too often thinks of himself as progressive and looks with scorn on the reactionary who asks whether the two courses could not just as well be combined.

Teachers have two responsibilities to their students: one, to give them enough information and vocational education to enable them to get a job and to hold it till they get rooted in a highly competitive world, and the other, to train them in methods of thinking and investigation to meet the demands of an ever-changing world—demands the details of which none can foresee. It is pretty certain, however, that the resources needed to meet the changes—the "challenges," to use current cant—will be the same in the future as in the past.

Inflation of the curriculum is not new. Many have seen it in the past; have watched some new animal brought into the college zoo, which was soon found to be a white elephant and later turned into a dun cow. But because new stalls had been built for these animals and they had acquired a group of expensive keepers, they were rarely, if ever, put back into the barnyard where they belong. A review of official correspondence connected with these courses and departments would usually show that an ambitious young man supported by some aggressive administrator had clearly proved that these courses were absolutely necessary for progress and had further shown that they would cost nothing—"involve no additional budgetary expense" is probably the correct academic phrase. And yet today one wonders why in the world they were ever established at all. The cow is never—well, hardly ever—returned to the barnyard, and animals that clutter up the zoo have been unusually prolific of late. Engineering in the undergraduate curriculum is becoming pocketed in smaller and smaller pigeonholes; it is time to consider the advantages of abandoning the roll-top for a flat-top desk in the educational world.

The most difficult and usually the most valuable element in the training of a student is the ability to synthesize—to put together the fragments of his knowledge into an intelligible picture. Specialization of the undergraduate curriculum goes in exactly the opposite direction. The student is allowed to synthesize either not at all or ineffectively; he has no guidance and no training in this type of work. The purpose should be to educate the student, not to inform him. This purpose cannot be served by the inflationary device of survey courses.

Some think this multiplication of courses is necessary for development of the research attitude among the younger men of the staff and among the undergraduates. Actually, all engineering is research if by research is meant the solution of a problem not previously encountered or the development of a new solution of an old problem. But engineering is not primarily the method of the organized research laboratory.

Very plausible arguments may be presented for including

all sorts of specialized courses in the curriculum if these two dogmas could be accepted: (a) that the universities should solve all the problems of the world in their own walls instead of training men who may help to solve them and (b) that differential specialization ever solved anything very effectively or that anything very reasonable ever came out of pure reason alone. Both dogmas are inacceptable. The actual production of any man, however productive he may be, is insignificant in comparison with the cumulative production of groups of men trained by some great teacher.

It is very desirable that the universities lead the thought of the people. Sometimes they do so. But on every campus are professors panting to catch up with the groups that they are trying to lead only to find that the group is one that got lost from the main body. As one looks back over the new courses of the past, it is apparent that they often represented digressions from the king's highway, not new roads to progress, and that their contribution to progress came after they had been brought back to the main road.

Specific digressions from the main objective of under-graduate training in engineering can be justified by one of the five reasons for drinking wine: "good friends, good wine, or being dry or fear you may be bye and bye, or any other reason why." Some learned this years ago when structural engineering became steel engineering, rigid frame engineering, masonry construction. Foundations went to college and came home high-hat, hydraulics returned from the

grand tour as fluid mechanics, and the whole group of in-
determinate structures has gone snobbish. All of them have
on some campuses scorned the older members of the family.
Must these new developments have a private suite with valet
and bath, or would they not do just as well if they sat down
at the family table sometimes to a meal of corned beef without caviar?

These observations are not based on any one institution
or group of them; they are representative. The main thesis
is that he who makes two courses grow where one grew before is presumably the enemy of progress in educational
method. Dr. James B. Conant refers to the "widespread feel-
ing that the separatist spirit of the past quarter of a century
has proceeded too far." This is true in the general field of
learning and it is becoming acutely true in engineering. The
function of universities is largely to produce men who by
becoming inquisitively rounded, by knowing their four w's
—why, what, where, when, the perpetual quadrivium—may
learn to see the world in its fullness. And so the business of
the universities is to train those men whose interests are
connected with the control and adaptation of natural forces
that they may become good engineers, whose pride and joy
and hope and salvation lie in the excellence of that which
they produce for the use and convenience of man.

To produce for the use and convenience of man they must
know something of that use and convenience as well as the
methods to be used in producing. The methods are called,

in the cant of the schools, "training in basic principles"; the knowledge of use and convenience, "broad training." The fight continues, and will continue from age to bewildered age, as to how to make the poor student both broad and deep, and arguments on it seem often to imply that the result must be the same for all students—and damn the time involved.

There is certainly a tendency in much current technical literature to enlarge the base. A certain college catalogue explains that the technical courses give the basic theories underlying the fundamental principles on which the science is founded. Now when you dig out under the foundation, you have quite a hole in the ground and to enlarge this basis you have to move a lot of dirt. The hole may be so deep that the student can never climb out to fresh air again or in such poor soil that the cofferdam of education caves in on him, to his great and permanent detriment. It is very hard to get students inured to having cofferdams cave in on them.

In the field of engineering the known basic principles are not very numerous; they are rather easy to state and to understand. The difficulty comes in applying them, and here a great deal of training is needed. No one can say how long this training must be. Many of the older men have not finished their education yet. But colleges can and do start students on this long road of training and can tell them something of the conditions, detours, narrow bridges, traffic signals.

Teachers are told that they must make their students

broad. And how shall that be done? By giving some more courses in sociology, economics, history, psychology, literature? Not unless there is some interest in them. All of these disciplines appear when engineering principles are applied to the planning of engineering works. If they do so appear, the interest is created and the student may then or later seek out books, courses, men who can aid him in satisfying that interest. Time can be found in the undergraduate course to let the student begin this search provided the courses in engineering create the interest.

Teaching is an art. It is not a science. A most disintegrating intellectual influence today is the idea that all human activities can be mastered by the methods of the physical sciences. As an art, teaching is necessarily individual; it must adapt itself both to him that gives and to him that takes as well as to the subject taught. It can be accomplished by lecture or by discussion. There are valuable courses for undergraduates that do not contain a single problem as assigned work, but some teachers use numerous assigned problems with success.

A novice asked Rafael with what he mixed his paints. The master replied, "With brains." A dean is said to have told his faculty, "I assume you are good teachers; your rating with me depends on your publications." The frankness of that dean is admirable, but everyone who daubs a canvas is not a Rafael.

Many people feel that teaching is just a job, like any

other job. In a sense that is true and it needs emphasis. A
teacher's first job is to teach, not to write or to do conventional research or to make speeches or to run errands on
academic or technical committees, but to teach. Do not misunderstand this; for a teacher, to keep his feet on the ground,
to keep in touch with the spirit of actual work as distinguished from the hothouse atmosphere of a school, must
serve on technical committees and attend conventions. It is
there that he is sometimes told bluntly that he does not know
what he is talking about, and most teachers need that badly,
for their intellectual mortality is deplorably high.

Not only must they be in contact with developments in
technical societies but they must also follow the relations
of those technical activities to other developments. In the
university and in the world outside the professors must attend many meetings, must talk to many people, must visit
many plants, must consult at many laboratories in order
that they may bring into the classroom a full vision, a fresh
outlook on the problems that they wish their students to
discuss. This all creates an environment in which men can
be taught to see America and American problems as a whole
and even to look beyond these.

Scholarship? Of course. How can the blind safely lead
the blind? The great teacher must know his field, must know
it in a peculiarly clear and vivid way. Then he will not only
be a thinker, but an original thinker.

Productivity? A teacher constantly trying to master his

field almost inevitably produces—research, books, articles, addresses. The by-product should be valuable, though much of it is not because so few academicians know when to use wastebaskets. The output will have value, if any, because of its quality and not because of its quantity. But all this does not affect the fundamental truth; the teacher's job is to teach.

Research? Oh, yes, that goes with scholarship. Hard and intelligent study of any field of knowledge inevitably leads to research, if by research is meant systematic investigation; in fact the distinction between scholarship and research is not clear. If the hard work is guided by the intellectual equivalent of fasting and prayer—really wanting to know, really caring enough about knowing to think hard—it will often be valuable research. But this is an incident to the teacher's work. He wants to know, not in order to be a "research man," but in order that he may teach well. That's his main work.

Teaching is an art. The teacher's job is to teach. What shall he teach? The amount taught is certainly not very important. Any well-trained man can take one or two books in almost any field and get from them over the week end more information than an undergraduate would acquire in a semester's course, and vastly more than he will remember —more information, that is, not more understanding. If the undergraduate has been well taught, he will know what part of this information is fundamental and what part ephemeral, what part is important and what incidental. Under a great master he will have formed some basis for critical judgment in the field.

What shall teachers teach? That is one of their great responsibilities, to determine what shall be taught, what to leave out, what to emphasize—especially what to leave out. That responsibility does not rest with the dean, certainly it does not rest with the student. It is so easy to give the student what he likes, to give a popular course. But there is no escape; the teacher carries the responsibility to decide what to emphasize, what to omit. It is his job to teach, and it may be added here, the student's job to learn. The student is there, please, to study how to do research; it is pretty bad to tell him he is already a research man.

The curriculum? One should be careful not to put over-emphasis on it, as far as teaching goes. It is usually revised every few years and the revision is often hailed as the beginning of a new era in education. But all the new developments could fit as well into the curricula of thirty years ago as they do into the more modern ones. The fact that revision of curricula for administrative purposes or for advertising purposes is often desirable is another matter.

How shall the teacher teach? By winning such affection that students will gladly follow where they are led or that their minds may flower to perfection in the glad sunlight of love and sympathy? This is fine, but, as the mathematicians say, it is neither necessary nor sufficient. This is a discussion of teaching, not of how to run an intellectual nursery. Some of the most popular teachers are mighty poor ones; some of the greatest teachers have not been generally loved. Students are usually fair and about as much in earnest as their teachers. They will follow the lead of the teacher who has mastered his subject and his art. They may not love him, nor need they do so. A good many so-called popular teachers achieve popularity by prostitution of their art—and students know it, but it is an easy way out.

Again, how shall he teach? There is no conclusive answer to that question. Uniformity of method is certainly the last thing to be desired. It is not necessary that all pictures of girls be Gibson girls; a few Mona Lisas are acceptable. There are many kinds of teachers, many fields of thought. Even in the same field of thought different men approach their subject by different paths; there are several approaches to even the most specialized subject. If these roads to understanding can be mapped without too much confusion—and often they cannot be—that is good.

And how can this great teacher be identified? Well, often they are not. Rafael was a great painter, but no one ever heard that it was because his dried paint had a high Brinnell number. This art, like other arts, must often be its own reward. It is inevitable that the administrative type of mind is usually quite different from the teaching type, that many excellent administrators have trouble in recognizing great teachers. It is also true that many educators are like the asylum inmate who explained why his friend could not be Napoleon. Sometimes everyone agrees on the greatness of some inspired teacher.

Teaching facilities—lantern slides, fine desks, handsome buildings? They are all right, but it may be observed that the really great teacher will, to use a homely phrase, teach in spite of "hell and high water." Great buildings and expensive laboratories can never make a great university; great teachers do.

There have been many good teachers and some really great ones, though these are scarcer than gold dollars. They have been of different kinds; some great graduate teachers, some gifted in the undergraduate field. One can recognize them by the vision, the inspiration, they give to the men they train. There are some who think that these rare great teachers are by far the most important men in the educational world.

Those who have taught for many years frequently observe advantages and limitations of various disciplines both in fact and in the hopes of opinionated partisans. But education is of a whole man for a whole world—humanities, urbanities, banalities. He who is either unable or unwilling to correlate the phases of intellectual experience is likely to contribute little to education—of himself or of others.

The professor spends relatively little time in actual contact with the student. In general, an undergraduate student does not spend more than a week or two of actual working time in the classroom in contact with any one professor. Put in that way, the statement is rather startling. Many, looking back, will recognize the tremendous influence some professor exercised in their growth and yet they rarely remember exactly what he taught. The professor acts in part as a catalyst—a material which assists in a reaction; after the reaction has produced a new material, the catalyst remains just as it was before, just as uninteresting but just as potent, ready to catalyze and repeat indefinitely. Students are educated through the catalytic action, if the professor has the personality to bring on the reaction. They are almost always in an intellectual environment which is conducive to growth, an environment of laboratories, museums, libraries, pictures and discussion groups. All of these are effective and should be easy to find. Universities should be able to say to their students, "Ask and ye shall receive, seek and ye shall find, knock and it shall be opened unto you." So professors want many things to create the intellectual environment. Sometimes an alumnus thinks they want too many things, but it may be found that the extent of a professor's wants are sometimes a measure of his value. The more energy he has, the more interest he has; the broader his interest, the more things he will want.

To reach the full development of their capacities, men pass through three stages. At first they use certain routines, formulas, fixed specifications. Anyone who professes to train engineers and turn them out into the cold world of facts without some discipline in the use of the standard procedures on which modern industry is based has not played fair. But the students go on and may expect to become junior executives. They are then in a position to revise, discard or invent routines for others to follow. The purpose of the mechanical brain in the evolution of modern industry has been very much the same as that of the assembly line in manufacture. Formalized procedures are set up for the guidance of men of less experience. Eventually, it is hoped that young men may reach a third stage and be able to take the scientific group, the economic group and the social group and put them together. Their problems then have ceased to be formal engineering problems and have become national problems, problems of industries, problems of the use and convenience of man.

Alumni are much inclined to have educators carefully prepare some part of this long road that young men are to follow and wish to have attention concentrated on the particular part of the road that they themselves are traveling at the time. Thus a young graduate of thirty often thinks that he should have had more technical details in his courses. At forty there is often the complaint that not enough attention was given to law and management, at fifty the alumnus wishes that he had studied more English or that he had read more of classical literature, at sixty he is usually grown up enough to recognize that colleges are dealing with young men of twenty and not old men of sixty and to realize that it is best to harmonize and give due attention to all stages of his career.

The Education of an Engineer

TO LIVE A FULL LIFE IN A BROAD WORLD

"Who through the blessing of Almighty God may be fitted for Publick employment."

IT IS customary to think of engineering as a part of a trilogy, pure science, applied science and engineering. It needs emphasis that this trilogy is only one of a triad of trilogies into which engineering fits. The first is pure science, applied science, engineering; the second is economic theory, finance and engineering; and the third is social relations, industrial relations, engineering. Many engineering problems are as closely allied to social problems as they are to pure science. The limitations of academic classifications are notorious. The workaday world does not fit into an academic department or into so-called fields of learning. It is the whole man who works, the whole community in which he lives, and it is the function of the university to look over and beyond its rather sterile classifications.

Mechanics, for instance, is a diamond of many facets and scintillates with different colors for the mathematician, the student of pure physics, the student of cosmic physics or the engineer. To nature it is undoubtedly the same mechanics, but it seems futile to think of it as a unit in the intellectual approach of the investigators. H. M. Westergaard wrote: "It should be remarked that the theory of elasticity is primarily physics, aimed at the understanding of matter. The development of the fundamental processes of theory, through the past one hundred years, has been the joint work of physicists, mathematicians and engineers. Applications to the molecular theory and the theory of sound have presented themselves. At the same time, applications to structural analysis have been a cause of continual contact with engineering. These practical applications to engineering have come into the foreground during more recent years."

A. E. H. Love explains: "The history of the mathematical theory of Elasticity shows clearly that the development of the theory has not been guided exclusively by considerations of its utility for technical Mechanics. Most of the men by whose researches it has been founded and shaped have been more interested in Natural Philosophy than in material progress, in trying to understand the world than in trying to make the world more comfortable. . . . Even in the more technical problems, such as the transmission of force and the resistance of bars and plates, attention has been directed, for the most part, rather to theoretical than to practical aspects of the questions. . . . The fact that much material progress is the indirect outcome of work done in this spirit is not without significance. The equally significant fact that most great advances in Natural Philosophy have been made by men who had a firsthand acquaintance with practical needs and experimental methods has often been emphasized; and, although the names of Green, Poisson, Cauchy show the rule is not without important exceptions, yet it is exemplified well in the history of our science."

Some engineers have studied the classics as well as the more customary engineering courses; have attended so-called "liberal arts" colleges. Here they were exposed to a curriculum largely dissociated from the problem of making a living. However, many of these engineers will admit that much of what they got from that curriculum has been the most practical training that they have had in engineering, though that curriculum never explained that blueprints were not made with white ink. Undergraduate work in engineering should provide men with the assurance that the college with its background of liberal education, untainted and uncorrupted by the desire for practical application, will provide engineering students with a background that will supplement and support their technical training.

In addition to necessary classical interests engineers need character and culture and charm—and so does every man. There is probably no surer road to the development of character than straight, hard, courageous thinking. As for social charm, the colleges have no monopoly: any student who will fairly compare himself or his classmates with men of equal mental endowments and social advantages who entered business directly from high school will realize this—a matter of common knowledge in the business world. Those who live without culture, without a knowledge and appreciation of the beautiful in the past and in the present, will only half

live; but it is a very common error to assume that cultured
men of eminence are great because of their culture, whereas
the truth is that they are cultured because of the cosmopol-
itan interest which helps to make them great.

If culture represents realization, appreciation and enjoy-
ment of the fullness of life, of all the material, mental,
aesthetic and spiritual factors that make up this world of
men, engineers are in a peculiarly favorable position to
achieve it. If they enter fully into the science and the
humanities involved in adapting natural forces to the use
and convenience of man—well, that is culture; then engineers live it and make it.

That is their privilege, to live life fully, to see the beginning and the end and the influence of their work; to know
the birth, growth, decay and rejuvenation of railways, the
changes in inland navigation, to work with architects,
lawyers, economists, statesmen, with materialists and humanists. Few men ever live life fully, but the opportunity
and the birthright are there for the engineer.

There is also an obligation. Engineering, of necessity,
profoundly affects culture. Engineers should not be in-
articulate. They need to tell others—not each other—how
they achieve their results, not the boring technical details
nor the mathematical processes that laymen misconceive.
Instead engineers must explain that their work results from
careful weighing of evidence, from examination of many
possible solutions, that only after judicious discussion of

past experience, present conditions and possible future developments are solutions accepted. It is important that men know that engineers do not build alone with concrete and steel or by formulas and charts, but more than anything else by faith, hope and charity—faith in their methods, their training, in the men with whom they work, faith in humanity, in the worth-whileness of life; hope that by use of these they may find men, money, materials and methods, not blind wishes but judicious hopes; charity that involves a sympathetic understanding of the human element and willingness to work within the limitations imposed by human weakness. Engineers should decline to undertake enterprises on unsupported faith or vague hope and few engineers have much toleration for undiscriminating charity.

This discussion of culture may excuse a digression to a fascinating side line of structural engineering. Nearly everyone is given to some hobby of collecting. Most undergraduates have a passion for collecting formulas and graduate students are much given to collecting all sorts of methods of analysis. Both varieties of bird's-egging are likely to become vicious, and engineers might try, as an outlet for such postage-stamp proclivities, an excursion in some field such as bridge collecting. By photographs and descriptions, accumulation of historic associations and of artistic detail, one can build up a museum which is not only of interest as a hobby but has value also as a background for professional work. It is a real pleasure to turn from the exact mathematics of analysis or the details of connections to a more general view of the function of bridge structures. A group of bridge pictures will enable the engineer to see bridges, not as formulas, but as studies in light and shadow.

A bridge must be structurally sound, correct in form, adequate in detail, of good materials properly used; but it should also fit into the landscape and with grace and dignity carry the roadway over from street to street or from hill to hill. The distinction between architect and engineer is quite recent and in bridge architecture it is almost impossible to enforce it. One who would design a beautiful bridge must have correct concepts of structural action; the artist must be something of an engineer, the engineer an artist and planner.

One of Maxfield Parrish's murals has an inscription in Gaelic. "Here's to the bridge that carries us over." That's what a bridge is for, to carry the roadway over, but it may do it in any one of many ways. The bridge is a part of the road-way, and also a part of the landscape and of the river or valley that it crosses. It must harmonize with its environ-ment; it must meet the spirit of its associates. In a park it may be a jolly little bridge, and play, as a little suspension bridge over the lake seems to play in the public gardens of Boston, but it must be a very serious-minded bridge where it is to carry a railway over a gorge. If it lives in pine forests, the bridge will perhaps want to be of timber and feel that it fits into the neighborhood, but rock gorges call for cut-stone masonry or concrete, and for huge spans the strength and grace of steel are used.

Paris, with all its fascination, center of art, ancient seat of learning, city of great vistas, of magnificent gardens, is also a city of beautiful bridges. Artist, architect and engineer find fascination along and between the banks of the Seine. Pont Alexandre, Pont de la Concorde, Pont Royal and all the bridges connecting the island with the banks fit gracefully and harmoniously into the magnificent vista from Notre Dame to the Trocadéro. At Paris, as elsewhere in Europe, the accumulation of beautiful bridges has been accomplished through long selection. The beautiful bridge is a bridge well designed; a bridge well designed is, in general, durable. As the years go by, it becomes part of the life and affections of the people, a part of a city, a focus for civic development. It captivates the fancy of artists and poets, and so endears itself that it is permitted to survive with small change as the years pass.

Europe has many examples of the quaint and the beautiful in bridge architecture in so far as Europeans have been able to preserve the best of their ancient bridges. However, their more recent bridge architecture is not superior to that in America. This may be seen in the newer bridges over the Seine and the Marne. At Château-Thierry, for example, the modern bridge of reinforced concrete seems mediocre and harmonizes little with the ancient buildings along the river or with the moldering castle on the heights; one feels a little sympathy for this new material forced into such ancient and distinguished company.

America today is developing excellent standards in bridge architecture. In the past American engineers have been so busy building bridges that they have sometimes forgotten that beauty as well as usefulness is an important property. But where a bridge has been "right," of materials that obviously fit into the community and of design that is structurally correct, American bridges have a dignity not surpassed in Europe.

Europe has few bridges that could be called large; the bridge over the Elbe at Hamburg and the Forth Bridge are among the few that by their size alone would attract attention in the American technical press. To these may be added a few over the Rhine and perhaps over the Danube. But some of their larger bridges fascinate by their curiousness rather than by their beauty. Forth squats spraddle-legged in the Firth like an antediluvian dinosaur, magnificent in size, but not distinguished in proportions. America is the home of the great bridge.

Bridges present one face to river travelers, another to those journeying by land, and a third to those who loiter by the parapets to fish or rest or dream. Much fine art has gone into the study of approaches, of pier forms, of details of balustrade. Each bridge has its own environment; it may be merely an extension of the street and be dominated by neighboring buildings, as is the case of Ponte S. Trinita; or it may itself dominate the view as does Risorgimento.

Fine bridges have a personality of their own. The Lars Anderson bridge in Cambridge, Massachusetts, charms because of its companionship with the river; the bridges of Venice are part of that glorious ensemble of renaissance architecture; James B. Eads's bridge has grace and strength of line in keeping with the dignity of the Father of Waters; the Charles Bridge over the Moldau at Prague fascinates with its Jew's Cross and other fine statuary; some bridges play in the park, some majestically span great rivers, but the bridges that impress themselves on the imagination always fit into their environment. Ponte Vecchio is charming over the Arno, Ponte di Rialto is part of the Grand Canal, but the Chicago River is another stream with another tempo.

Europe seems to have loved its rivers and its bridges more than Americans have loved theirs. The embankments of the Seine, of the Thames, of the Tiber, of the Alster Basin all reflect the fact that in Europe the riverbanks have been more fully developed to charm and rest those who pause to enjoy the hospitality of the bridges. Americans have appreciated the rivers in their own cities much less than they should have done and less perhaps than they will in the future. Boston has done marvelously with her Charles River Basin; Chicago is changing its river from a great sewer to a ribbon of restfulness; Pittsburgh is discovering the waterways whose junction made her a trading post; and Indianapolis has found that the White may be a thing of beauty as well as a flood maker.

Engineering then is not merely mathematical science. It must be approached with a sense of proportion and aesthetics. In so far as engineers deal with facts that can be measured they use mathematics to combine these facts and to deduce conclusions. But often the facts are not subject to exact measurement or else the combinations are of facts that are incommensurable. There is no special difficulty in comparing two distances, but many a car driver argues with his family over the relative importance of miles of driving and mountain scenery. The work of the engineer deals with human customs as well as material facts. Municipal engineering furnishes familiar examples—the relative importance of parks and parking space, of convenient neighborhood stores and zoned residential districts, of subways and sunlight. The importance of such problems emphasizes the need— the very practical need—of the engineer for a knowledge of history and literature, a key to how the mind has worked in the past or will act in a new environment.

An important duty of teachers is to force students repeatedly back into the field of reality and even more to teach them to force themselves back into reality. Some seniors forget that the laws of mechanics make them fall and bump their heads, that British thermal units scald their fingers, that energy may kill. In fact, there is scarcely any absurdity to which seniors will not agree if it is presented with enough Greek letters and integral signs. Some of them seem to have lost completely any will to check their conclusions with everyday reality.

Many teachers attempt to overcome this difficulty in the laboratory. But the laboratory model is not the same as the structure in the field and is often far from it. An engineer once described a certain test for materials as a test applied to a material in order to determine whether or not that material would pass that test. Often the sense of reality seems to decrease with the elaborateness of equipment and finally disappears completely from laboratory work.

The usefulness of technical devices should be measured largely by the degree to which they are based upon the world of reality and experience. Standardized tests, symbols, formulas and technical terms should not be permitted, as they often do, to supplant reality in college courses.

It is easy, as many engineers know, to go too far in the pursuit of this elusive reality in college; it is a common failing to assume that if students see enough bridges and pictures of bridges they really do not need to know much about the analysis of them. "Practical" courses dealing with how-it-is-done, to the exclusion of why-it-is-done and of how-it-might-be-done, are largely a waste of time.

What does happen in the structure is frequently not as important as what may happen. What may happen may never happen; probability is a concept, not a reality. This elusive illusion of reality, however, is not the difficulty that undergraduate students encounter. They have lost even the illusion of reality. If they are asked to draw a structure deflected under loads, they will draw a wiggly line which any bumpkin on a springboard should know is not even approximately correct. They compute a negative reaction on a cantilever beam without interest in the meaning of the negative sign. This seems to show that the schooling process has destroyed in students something of great value; certainly either the student or the course has lost something vital.

Good engineers have a vivid sense of reality. The judgment of the engineer who has it is worth much more than computations by men in whom it is feebly developed. There is no more important question for teachers of engineering to think about than how best to develop it.

Engineers should be persistent and also judicious in their wants. There is here a combination of two things, a nice adjustment of individualism with regimentation; the one permits the individual to express himself in his work, the other precludes extravagant and unprofitable experimentation. This is not meant as a discussion of the desirability of regimented humanity or of complete freedom for everyone at all times. All want more individualism in design and more standardization in detail, the two are not discordant and incompatible but are desirable and coordinate. There is also a persistent interest in use and convenience. Engineers are primarily conscious of and intimately concerned with the consequences, social and political, of the works that they have in hand. If there is any blame here it rests with promoters and financiers and not with engineers.

Another important element of intellectual training is coordination of analysis and synthesis. The present age almost certainly tends to carry analysis too far, and engineering schools in most cases have favored this tendency. The ultimate objective for engineering is planning and building. The function of analysis is incidental to this, but it serves as a guide to the ultimate carrying through of the plan.

Most important in this picture of design is the sense of scale. Some men never seem to get it—the ability to recognize quickly that certain phenomena, certain stresses are important and others not; the significant ability to put first things first; the ability to weigh the consequences of failure and adapt a factor of safety to the probability and to the consequences.

"Scholarship," "research," "productive investigation," are often the last refuge of academic charlatans. Scholarship for engineers means first, that they know accurately what they are talking about. This implies, as a fundamental, complete honesty, as well as high intelligence and a lot of lonely hard work. It implies, as corollaries, accuracy of quotation, accuracy and precision of documentation. Scholarship for real accomplishment and love of culture are high ideals, but it is ever true that men are broad and cultured because they are great and did not achieve greatness by pursuing culture.

Engineers are often so anxious to do that they are not very

systematic in knowing. Their papers are too often poorly documented, their quotations are too often secondhand. Now perhaps engineers do not need to be good scholars, but they do need to pay more attention to some established rules of scholarship. These rules have been used more systematically in the fields that have a long bookish heritage than in engineering, where many of the important thoughts and facts never get into print at all. It may be very annoying to read that certain statements are supported by tests without any suggestion as to where the test records may be found; it is amusing to find an author relying on the authority of an author who quotes another; it is disgusting to find references to sources not available to the author or in a language that the author cannot read. These cases are violations of plain rules of intellectual honesty. However, some authors apparently do not recognize them as such. Intellectual honesty implies an intellectual tradition comparable with the material tradition back of material honesty.

Seniors should be made to realize that the university is a place to get into as much intellectual trouble as possible, a place to make mistakes, many mistakes, and to rectify them. They usually think they know what this means but actually they do not. It is not the quantity of their mistakes that should be improved—they make enough of them—rather it is the quality. They do not get into any new troubles; there is rarely the charm of individuality or originality in their errors because they lack the courage to try intellectual experiments. Juniors are taught to get one reaction of a beam on two supports by taking moments about one support. It never occurs to them to take moments about two other points to get the two reactions. That would be an experiment and if they tried it they would see why that is not done and would find out a trick in thinking that has wide applications in engineering.

Does the engineering curriculum train for leadership? This jargon of leadership is mostly nonsense. The university can take men of reasonable health, ambition, character and intellect, and put them in an environment in which they will learn something of leadership, of its realities and its failures. Clearly all men cannot lead. There is a great deal of misunderstanding of leadership; a cynic has defined an executive as one who assumes all prerogatives and avoids all responsibilities. Engineering training can provide two things that are somewhat difficult to get except in similar fields of thought: ability to observe and ability to interpret important phenomena of nature with some measure of accuracy. How hard does the wind blow? How much will it rain next year? What is the probability of flood? What is the force of storm waves? What is the strength of timber or stone or brick? The value of being able to observe and critically interpret is greatly enhanced if students learn to arrange their information in a usable way. They can be taught the difference between a fact and what someone claims or hopes is a fact. Much that must be taught to those who are trying to become

engineers consists of definitions of terms, important principles in algebra and geometry, arrangement of computations and the language of drawing. Much of it consists of
teaching the language. In addition they should be given certain information about materials and sometimes about
methods of construction.

Engineering changes—the character of the literature, the
problems, the types of structures and machines—and there is
pressure to modify radically the training given to engineers.
This pressure is especially strong from three groups: the
humanitarians, the research laboratories, and the graduate
schools.

Engineering is an old art. It has always demanded ability
to weigh evidence, to draw common-sense conclusions, to
work out a simple and satisfactory synthesis and then see that
the synthesis can be carried out. Because the art constantly
adapts itself to the use and convenience of man and because
there are changes in this use and convenience, the emphasis
in the development of the art varies from generation to
generation, from decade to decade. It forever adapts itself
to change and yet the more it changes the more it remains
forever the same—and this must be true of the education of
engineers. If the young men of America learn the importance
of judicious wanting, are trained in digesting evidence and
learn to study the customs and convenience of mankind,
they will be able to adapt themselves to new problems and
new materials.

Should students be trained to be conservative? Here they can learn that there are laws and forces that they cannot modify. Should they be taught that they can't get something for nothing? All engineering design emphasizes that, has always been based on that. Should they be given resistance to propaganda, to directed statistics?

Teachers are important, very important. They cannot completely ruin a good man nor can they make a barrel out of a bunghole, but they can accomplish much in either direction. Undoubtedly they can be invaluable in indicating those methods of thought and study that are commonly unprofitable or actually harmful. They can help the man to grasp the idea that engineering is not a branch of mathematics, though mathematics is useful to the engineer; they can discourage purely speculative studies that have no purpose; they can get men to realize that engineers ask "What of it?" as quickly as "What is it?"

Perhaps the most valuable training that the college can give is in the use of books. Few students know how to use them. Few can realize the hesitation with which a discriminating author selects his material or how reluctant he would feel to say that all this is to be swallowed and that's all there is to the subject. The information in books is secondhand to the student and secondhand information carries the same dangers of disease germs as secondhand clothes. The student must be made to proceed cautiously before accepting such an offering.

Critics of the education given to those who wish to become engineers are not helpful when they start with the thesis: "Sugar and spice and all things nice, that's what humanitarians are made of; rats and snails and puppy dogs' tails, that's what engineers are made of." Often their criticism is based on conclusions determined by misapplying methods of thought that they got from engineers. They condemn engineering education because of inconclusive miscellaneous information and "statistical data" that the engineer would at once reject because of loose definition, inaccurate collection, confused classification.

Some today would prefer engineers trained as psychologists, sociologists, economists, politicians—each exclusively and each under the guise of adapting engineers to the world in which they live. Others suggest that young engineers be made into research specialists, experimenters, physicists. Still others urge that they should be given character and common sense and conservatism. Some seem to mistake the teacher for the Almighty.

As a consequence, the curriculum of engineering is being pulled in different directions by followers of different philosophies. One group would emphasize research, another the creative elements, and others would include so much of the general knowledge of other disciplines as to leave little room for essential training. The demands are for general surveys of civilization, of the continuing elaboration of mathematical mechanics, of laboratory training in technique, of design, of details. This conflict of demands is wholesome; life, if full and dynamic, is like that. All these things or at least some of them should be—and usually are —in the curriculum, but it does not at all follow that they should be there as separate courses.

Setting up new courses does not and never will meet old needs. Men can learn a good deal of sociology and economics and political economy in connection with a course in sanitation or highway engineering or in almost any other traditional course in engineering.

If somewhere engineering—including detailed design, analysis, synthesis—is taught, all these matters become involved; what is done in four or five or six years is important only as it trains for the remaining thirty or forty years of useful life. There is sometimes cause to fear that scientific technique, the proud servant of the engineering arts, is trying to swallow its master. In regard to the multiplication of courses in the curriculum, it has been said that some seem to overlook the invention of Gutenberg.

Men who guide the training of engineers should keep the long-range objective ever in view and remember that there are several stages in the growth of engineers if they are so fortunate as to complete their growth. They start, as Shakespeare suggested, in the nurse's arms—the protective arms of Alma Mater—with complete excremental analogy. After graduation they get jobs doing a fairly specific thing in a pretty definite way. During those first few months no one expects from them great constructive thought. They are asked to carry out a few fairly well-defined procedures. Before many years they pass into another stage in which they put together information from several sources, bring into their problem the human values that affect it. Later still they begin to create the problem themselves and so they grow, from young engineers to managers of industry or of great projects and are then perhaps not called engineers at all.

Probably few doubt the importance of proper training for men who will plan or be associated with the planning of engineering works in America. America is the land of great engineers. These men must know how to use science to further the welfare of men, though they need not necessarily be scientists in a narrow sense, nor academic specialists either in defining welfare or classifying men. Their work is outside ivy-covered walls, where the world's work is ultimately done and where on a broad level it is best thought out.

Science and system, law and custom, men and manners. The universities will not correlate these, but the universities have a great obligation, a great opportunity to show students that later they themselves must try this correlation and that the sooner they start, the better. The engineering curriculum is full of science and system; more of it cloaked as sociology or statistics or formal mathematics does not help. Examples of constructions, machines and processes not well adapted to people, of customs and usages that conflict with mechanical progress teach the man that unchanging nature must be directed for changing life. "The art of directing the great sources of power in nature for the use and convenience of man"—art, not merely science; directing, not merely observing; convenience, as well as use; man imperfect and apparently not perfectible but very much alive. This definition is still good; it is really too bad that it is so often forgotten.

Minarets Above the Ivy

GRADUATE STUDIES, DISSERTATIONS,
RESEARCH

"Listen with credulity to the whispers of fancy and pursue with
eagerness the phantom of hope."

THE ENGINEERING curriculum for undergraduates must be changed from time to time not so much to keep it up to date—up to what date?—but to keep it alive, to keep it out of the museums. The more it changes the more it will remain the same, like the wind and the waves and the mud. In recent years the undergraduate curriculum in engineering has been subjected to an influence that scarcely existed thirty years ago, the graduate curriculum. Although some are still skeptical of its usefulness, it's here to stay. It can be a powerful stimulant to the vitality of the undergraduate curriculum but, like most stimulants, it is dangerous. Forty years ago the graduate curriculum was going to rescue the undergraduate curriculum from inanition in the field of liberal arts and many have seen it throttle the old girl while pretending to wake her from her sleep.

Graduate study is relatively new in engineering and for this reason the profession has a chance to save itself from some evils observed in older fields of study. Some successful engineering organizations consider a master's degree so important that it almost becomes a necessity for promotion; they even ask for men with doctor's degrees. More and more of the inquiries for teachers specify the master's degree. Much of the material in the transactions of various societies is a rearrangement of theses submitted for higher degrees. The productivity of the large graduate schools is enormous in volume. Very little of it is worth publication, but this does not mean at all that it did not serve its purpose.

It is not fair to feed students entirely on secondhand information. College instructors should be able sometimes to tell their students that such and such is true because they saw it in their own laboratory. Students must be brought close to the original source of knowledge. It is staggering at times to realize that some men have much schooling without ever knowing that there is such a thing as an original source of knowledge. Students should be taught not to jump at conclusions because someone wishes them to do so, not to accept too naïvely all the test data presented, not to try blindly to do what cannot be done. Sometimes this can be accomplished by having them review articles by great authorities that happen to contain flaws easily identified—the bigger the authority the better the lesson. American colleges carry a moral responsibility not to live on begged or borrowed brainwork, but to pay their way in the intellectual world in any field of learning that they sponsor. They cannot each live exclusively on the intellectual product of another institution. At the same time great universities owe it to the country to contribute collectively, through their research, an accumulation of knowledge in order that the present may pay its obligation to the past. This is an ambitious program, but it has been done before and it can be done again.

Graduate study in engineering should be a part of the general process of education, the purpose of which is to prepare a whole man to live a full life in a whole world. This statement is too general to be very useful in formulating curricula or setting standards for degrees, but it is important not to forget the ideal. There is an important difference between undergraduate and graduate students. The difference is not in their over-all objectives, but rather in the fact that undergraduates must be led while graduate students should begin to lead, so to speak, should be encouraged to develop personal responsibility. At times study on the graduate level should proceed much less rapidly than on the undergraduate level because time must be taken to examine critically all assumptions; at other times it should proceed much more rapidly because the undergraduate studies already completed have got some details out of the way.

In undergraduate work there is time to give individual training only superficially, if at all; in graduate work students are put more and more on their own responsibility until finally they have some idea of what is meant by accuracy of statement and have learned some discriminating humility. Too much emphasis on ignorance produces men who can argue all sides of questions and yet not settle anything even tentatively; too great elaboration of knowledge produces a dangerously erudite and clumsy pedant. These extremes can be—must be—reconciled.

It has been said that "In research, there are no standards." This comes near the truth in graduate study; but here again "All generalizations are false." The general objectives of graduate study are to indicate the broad scope of practical knowledge, to teach the importance of accuracy and precision in securing evidence, to give practice in presenting conclusions, to develop intellectual imagination, courage and honesty. To do this requires a well-balanced curriculum and a faculty presenting some diversity of personality, experience and outlook. In graduate training, as in a well-chosen library, "much should be tasted, a part may be swallowed, some must be thoroughly chewed and digested."

Graduate study in engineering may be considered from the points of view of the general public, of the institution concerned and its faculty and of the individual student. These are intimately related and probably no one of them is considered exclusively by any American faculty; from any of these points of view it is in the end the trained student that is valuable to the world.

Certainly graduate study in all fields should be sanely directed for the interest of the community; how large is this community—state, national, international—depends upon the institution and the nature of the study.

For the institution, graduate work performs several related functions. It forces the faculty to re-examine critically the fundamentals of knowledge, since no honest graduate teacher can sleep long. Professors of engineering are obliged to keep themselves informed on the progress of research in their field and should be constantly doing some sort of original investigation—experimental, analytical or interpretative—if they are to do good teaching. But the teacher must use discrimination in presenting the results of these researches to the students; the university must be a filter, not a spout, lest it drown instead of educate.

Graduate study also furnishes a relatively uninhibited field for experiment in education where methods of thought and instruction may be worked out for use in undergraduate courses; it raises the general level of thought in the academic life of the university. The investigations may and sometimes do enhance the reputation of the institution in the scholarly and professional world, though this aspect of graduate study is much overstressed; it is generally the by-products of training or improvement of technique that are valuable.

Graduate study in engineering should be based on the theory that a man is best trained who is best able to combine an intimate and critical ability in one of the many aspects of the field with a full appreciation of how broad that field is.

It is perfectly true that men cannot know a little about many things until they know much about some one field. Thus, one of the values of graduate study is that students look rather deeply into a subject in which they are interested instead of spreading their energies superficially over many fields, as some unfortunately wish to do. The value lies in this, that if the students are really of graduate caliber, really capable of thinking out new problems in new ways, by looking deeply into the methods of thinking in one phase of engineering, they have looked deeply into many phases of engineering. For instance, structural engineering may be chosen as the field of concentration. In this field, the data —from analysis, from the laboratories, from the literature of experience—have been quite accurately classified. Here a careful study can be made of methods of collecting and weighing evidence, of interpreting the evidence and of correlating the technical evidence with the data of legal, economic, governmental and sociological usage.

Obviously, this does not by any means suggest that engineering is restricted to the field of structural works, but it does mean that methods of correlation and synthesis studied in one field may be readily extended to other fields.

To avoid narrow specialization, a general course in planning may be closely correlated with the work in structural engineering and in mechanics. This course should cover planning for the use of land in cities or in highway and railway systems and the terminals of marine, motor, railway or air transportation and designs for the use and control of water for water supply and sewage disposal, for flood protection and the control of rivers, for the development of hydraulic power.

Courses dealing with structural design are usually concerned chiefly with collecting, correlating and interpreting data; to use these data constructively involves much imagination. Such imagination may be developed by planning projects in engineering involving the layout of cities and of their sanitary, transportation, industrial and power facilities. Such studies should consider the regions in which cities lie so that the natural assets of states, littorals, districts and of the nation may be fully and effectively brought to the advantage of the people. Structures, in the broadest use of the term, are one of the most important factors in the development of such facilities.

Training in laboratory technique and in the interpretation of data from laboratories is essential for judging the value of evidence to be used in deciding engineering problems. Such training may be secured in studying the common but variable material, earth. A course in foundation engineering may be highly successful for general engineering education.

A course in research methods may offer wide opportunities for specialization to those who wish to follow individual interests. It is a common fault of schools that they concern themselves too much with unusual erudite problems; they insist on experiments with a capital E to the neglect of experience with a little e, they pursue research with a large R to the exclusion of thinking with a little t. Charles F. Kettering, one of the greatest of industrial research men, is quoted:

"All the money in the world and all the people in the world can't solve a problem unless someone knows how. Problems are solved in some fellow's head. They are not solved in a laboratory at all. It does take an awful lot of effort to get a perfectly obvious thing organized in a man's head." And again, "The only reason you try an experiment is to cultivate your own thinking. You say an experiment failed. That is just your alibi. It was your thinking that failed."

But this does not affect the fact that beginners would do well not to think of themselves as specialists. In doing so they handicap themselves at the beginning of the race. Specialists suffer a good many handicaps. If they will look beyond their specialty, they will find many new ideas or improved points of view developing in other fields, and if they have imagination, they will see that with proper modification these ideas have much value in their own field.

But engineering deals with nature—not with its description but with its control. Nature does not conveniently split up into little pockets so that a man can say, "I am going for a walk and I shall walk on rock but not on clay or concrete or roads or bridges." The problems of railway roadbeds are not entirely unlike problems of building foundations; engineers must note the similarities without neglecting the important differences.

Graduate students are often quite vague about possibilities, objectives and methods in thesis writing. A thesis— some catalogues offensively call the doctor's thesis a "dissertation"—does not, or at least should not, differ from documents written every day by well-trained men dealing with practical affairs. To prepare these the students must be able to state the problem that they propose to study, to assemble good pertinent evidence; interpret, present and sum up this evidence. Students should be given the opportunity to proceed toward these ends largely on their own initiative and responsibility.

A good deal of time should be spent in exploring the field of investigation, finding out what has been done, what should be done, what can be done. This time is by no means wasted, for upon its judicious use is likely to depend the later development of the topic, but it is not immediately productive. Collection of data in laboratory, library or study is the next step. This is often the easiest part of the work, for if the students know what they want and where it is, it is usually not difficult to go and get it. The great difficulty, of course, is to know what to want. In spite of a popular aphorism, properly asking questions is usually more difficult than correctly answering them.

It seems so simple but is really difficult to state a question in engineering in such form that it is possible to investigate it. One of the most difficult things to teach graduate students is the great value of men who ask significant questions in such form that they can be investigated by available techniques. The list of titles of graduate theses is pretty drab to an experienced man; this is probably unavoidable, for beginners must learn to toddle before they stride. Nevertheless, graduate students should learn that the conventional question is often not the important question at all.

To collect the evidence involves knowledge of the literature of the field and of methods that might furnish data; analysis and synthesis, mathematics, experiment, observation, common sense are all important, and data may either be original or from records. Relative importance and dependability of sources vary with the fields of study and the nature of the problem. To appraise the evidence intelligently involves knowledge of the dependability of the sources and requires much independence and courage in forming judgments; it requires a scholarly accuracy that comes only from arduous training.

First work with graduate students should be to try to get them to develop honesty. This doesn't come naturally; it takes lots of work and training. New men will state quite definitely some technical generalization. If they are asked how they know that, they'll reply that all the books say so; all right, then, if they are depending on the books, they'd better say so. Later they will report that certain researches showed this. Much later they are prepared to say that they looked up reports of the tests, that the author claims that the generalization seems to be safe—not necessarily exact —and that his data seem to bear out the conclusion. Of course life's too short to do this sort of thing with every detail of the day. But in a specialty, where others are depending on accuracy, it becomes a duty. Many men not only do not try to do it, but do not know that it is possible or desirable to do it.

To arrange and digest evidence demands practice, assisted by study of good reports as models. The manner of arrangement depends on the scale of training and of ability, for almost anyone can collect some data but it requires skill to put them in proper or usable order. Moreover, some of these data should certainly be discarded or discounted or given relatively less importance. One of the most useful tools available to research men is a wastebasket. Unfortunately, too few have the courage to say frankly that something they dug up with much patience and effort has turned out on careful study to be either not very dependable or not very valuable or is irrelevant to the immediate purpose.

The data, after being collected and classified, must be presented. This is not so simple as it seems to the beginner. Presentation of the evidence requires a mastery of technique and of good manners and good taste in presentation. Engineers use four methods to present evidence: graphical— drawings, pictures, sketches; statistical—charts, tables, pictorial charts; symbolic—mathematics in the broader sense of the term; verbal. In using these they must exercise good sense. Drawings or charts that are not readily comprehended show incompetence; bad spelling or diction or punctuation, labored style, lack of unity or coherence or emphasis are serious defects; documentation and general editing should be uniform and adhere to some reasonable standard; brevity is always desirable.

In technical writing every word has a rather definite meaning, and even if the words are used with these meanings it is commonly a problem to make the idea clear to the reader. Sentences should have both subjects and predicates, the sources of information should be clearly indicated and in general the rules of good style, so far as they are a matter of rules, should be observed. In any case the purpose is to present the evidence, the data, the "facts" clearly, briefly and simply. Needless technical terms usually confuse and seldom impress; it is much better to write in a natural and unaffected way. Many students seem to think that some stilted form of words is essential to scholarship, that affectation will cover inaccuracy or that elaborated formalism is a valuable substitute for graceful simplicity. Some might even think that the introduction to a thesis on suspension bridges would be fine if it began: "When our simian ancestors first descended from their arboreal haunts, the pendant draperies of the luscious vine, so familiar to their parents, offered a mode of transportation over otherwise impassable intervals."

Most students think that the difficulty is to collect the evidence; interpretation and presentation are too often left, in spite of all hopes and prayers of advisers, as an easy week-end chore. To interpret involves much use of the imagination, a well-developed sense of proportion as to the relative importance of the sources, knowledge of the logic of the field of study and of the general fields of science pure and applied; it also requires fast and prayer.

Interpretation of the evidence is always difficult. Correct and gifted interpretation represents the highest attainment of the scholar. No one can avoid all blunders; but even a beginner should be able to avoid some. A frequent error is the attempt to draw too many conclusions. Often the data are inadequate to draw any conclusion; to show this clearly may be a valuable contribution to learning, much more so than to seduce the reader into some conclusion not warranted. Not enough use is made of the simple statement, "I don't know."

Clearly the summing up is important; students will learn more and more during the next forty years how difficult it is. A brief and clear summary may be the crowning glory of a good engineering report and a good summary is very hard to write.

The main purpose of the thesis is the training that students get from this work—collecting data, either their own or that published by others, evaluating these data and arranging them, weighing the evidence and indicating the probability and importance of the conclusions, and presenting all this in a way convenient and useful to the reader. Now this is the essence of all engineering reports, it is a professional attainment of great importance to anyone who is to be valuable as a leader either in engineering or in the many fields, often not designated as engineering, into which engineering may lead.

There is much talk of "research," of "original contribution" and of the "progress of science." That is excellent, provided these terms are sanely interpreted. It is perfectly true that a compilation made without judgment or discrimination should not be presented as a graduate thesis. On the other hand, truly original contributions in any field are rare in any generation if by that is meant that the originality presents a quite new way of thinking about the world and its affairs. There is little originality in the solution of a problem of stress analysis the equations for which have not previously been written in the special form used, but which result from manipulation of Lagrange's equations by methods pretty well standardized. At the same time this may be a very good and even a valuable piece of work and, if properly presented, may constitute a very acceptable thesis.

Theses may be experimental or analytical, using the latter term as the equivalent of "mathematical"; they may be bibliographical—some rarely valuable work has been done by listing, rating and judiciously classifying existing knowledge. They may be what is best called "synthetic," in that they collect conflicting data from many sources and try to "weigh the evidence" and present the basis on which it is weighed. Or a thesis may be a design; if so the evidence bearing on the strength or other physical characteristics of the proposed structure must be weighed, and this evidence correlated with the intangibles—"usefulness," "convenience," "social value"—to give a worth-while synthesis.

Students rarely weigh properly the relative difficulties of these types of work. A paper that discusses a "broad" topic may seem easy to write but only a master can present a paper of true value on such a topic. The mathematical thesis in structural engineering seems to many the highest attainment of scholarly effort. However, it is common because it can be satisfactorily produced with such training as can be given the inexperienced student in a classroom. The same is true of much experimental work.

What all this comes to is that choosing a subject and writing a thesis are pretty much a matter of common sense. Students should do a useful piece of work on some subject that interests them and show that they can use the tools that have been presented to them for such work. They should get some fun out of it, and so should the men who direct the work. Unless the work develops in an unusual way and really opens up a very specialized life interest, the thesis is something to be written as a part of a man's training and then forgotten. Too many men try to go on with them in later years when there are other things calling for their attention; too many of them are published in spite of wastebaskets.

The time involved in various phases of the work is important. Perhaps a general estimate might divide the time nearly equally between definition of the question or purpose of the thesis, collection of the data, study of the data and writing the thesis. Unfortunately many students get their schedule badly off balance, especially by underestimating the time needed for the writing. A good plan is often to try, if possible, to carry on the different operations simultaneously.

A difficult problem is the distinction that must be made between theses for the master's and for the doctor's degree. Most graduate students in engineering are candidates for the master's degree. An increasingly large number, however, want the doctor's degree; whether this is good has nothing to do with the facts. If the trend is to continue—and it probably will—a new philosophy for this type of training should be evolved. Many of the candidates for the doctorate have planned to go right into teaching. This is almost certainly bad, for one who is to become a teacher of engineering should be trained primarily to be an engineer, and association with the profession outside of the ivory towers of learning is absolutely essential.

Candidates for the doctorate should come up with some such attitude as this: "I am interested in this field of study and I am pretty sure that I can contribute to it something useful enough to justify the special time and attention that my training will require." A candidate must not be urged or even encouraged, in general, to undertake the doctorate. The profession does not want or need many doctors. Their training, if at all well done, requires individual attention and is necessarily expensive. Candidates must satisfy the committee that they are interested and can contribute.

Men studying for the doctorate gain valuable experience through two oral examinations; and at the same time these interviews serve a useful purpose for the examining committee. Both examinations for the doctorate impose a severe burden on the self-confidence of the candidates. Of the two, the preliminary should be the more inquisitorial. If candidates cannot clearly explain and defend their thesis in the final examination, they are not authorities nor are they likely ever to become authorities on any subject; the familiarity of the candidate with his subject in this examination should make him master of the situation.

If the preliminary examination tests the information acquired by the candidates in individual courses, it is a waste of time; knowledge of individual courses should have been tested in the courses themselves. This examination should be a test of the quality of the mind, of each man's method of thought in the field studied, of the genuineness of his interest in this field of study rather than in a curriculum and in a degree.

The proper conduct and therefore the usefulness of both examinations presupposes a competent, honest and courageous examining committee. Members of an examining committee are not very useful unless they are able to judge whether the candidate can think clearly. Members must be really trying to get at the facts, not merely seeking to justify a predetermined conclusion. To decline to pass a candidate is one of the most unpleasant duties that they may have to perform; they are, however, employed to form and state an honest judgment.

Some candidates become confused more readily than others. Then that's that; these men are to become authorities. The committee should be human, and most of them personally know the candidate. They should and usually can get below the superficial evidence and find out how much of the confusion is nervousness and how much is fundamental uncertainty.

There is to the candidates themselves a great deal of value in these examinations. They know that they cannot pass a course and then forget all about it. They must organize and digest their knowledge; they must correlate the test data from one source with the analytical theory of another. They are to be examined not on courses but on a field, and if there are gaps between courses they must have filled these by reading or by conference.

There is as yet a good deal of prestige attached to the doctorate. There is no more important academic responsibility than passing upon candidates; ambition to increase numbers of graduates or to be kind to aspiring young men must not blind committee members to that responsibility or tempt them to put the stamp of distinction on mediocrity.

For Man's Use of God's Gifts

CONCEPTS OF ENGINEERING ART

"All the days of the earth, seedtime and harvest, cold and heat, summer and winter, night and day, shall not cease."

THE CONSTANT and insistent need that engineers feel for any scrap of fact from which they may predict natural phenomena tends to develop a hunger for anything that even resembles a fact. This in turn may lead to a wolfish and gluttonous attitude, a gobbling up of every statement or opinion, figure or formula, indiscriminately and incessantly. The result is often intellectual autointoxication from "hunks and gobs" of unselected, undigested and indigestible material.

Rather engineers need to select their mental diet carefully and when they go a-fishing after facts they want a fish fry and not a chowder. Their fishing trips are often long and arduous and it is important that they take along only the simplest and most useful equipment; complicated toys, however beautiful, are to be avoided on these mental journeys. Definitions of terms are like the names of towns along the way, mathematical relations make a sturdy canoe to bear them and desire for engineering facts drives them on. At last they find their country, a land of lakes and rivers teeming with fish—facts of nature borne on by the unceasing current of natural phenomena, all sorts of facts, some use-

ful and some useless to them. And they spread their nets
and catch these fish and select what they want and use them.
And later they often tell about it after the manner of all
fishermen.

The net that catches mental fish is made of questions bearing on the subject studied. Hence, men trained in collecting
information begin first by collecting questions rather than
by collecting data. Indeed men's knowledge of a subject can
be measured better by the questions that they ask than by
the answers that they give; there is no surer mark of
ignorance than the assurance of complete knowledge. When
a subject is first studied, there are few questions; the mesh
of the net is large and important facts slip through unnoticed.
But if the student is awake each new fact adds new questions and, as the data are reviewed, new facts are perceived
and held fast in the mesh. At first the net is not very well made
and at this stage it is not always best to get a great many
facts, for the net cannot hold a large number of fish even
if it catches them. But if the threads are made stronger, if
the questions become more clear and definite as the study
proceeds, the net will eventually hang each little fact by its
gills. Then all the trout or perch or catfish can be strung on
separate strings and eventually put in the frying pan of design. If the net is not allowed to rot but is turned over in the
sun occasionally, it's all ready for another fish fry some
other day.

Of course, there are other ways to have fish fries. One

way is to dynamite a pond; that's "messy" and ruins the technique of the fisherman. Or several barrels of assorted fish can be bought and the fishermen can see how they like them. The trouble with this procedure is that the facts may be spoiled if got from an undependable person. Or you can go to a restaurant; but this is a discussion of how to be an engineer, not how to use handbooks.

To drop this metaphor, these last three ways of having a fish fry correspond in reverse order to three definite human tendencies of our minds, all based on the same motive. They may lead—and often do—to mental ailments, the pathologies of which are distinctive and important. Most people will go to any amount of trouble, effort and inconvenience to avoid the supreme agony of concentrated thought; and yet they know that no trouble or effort or inconvenience can avoid the final need of it. And so from fear of mental exercise they become exposed to the maladies of formularitis, translatitis and experimentalitis.

Formularitis appears at every age, in every clime, in every field of thought. It attempts to reduce cases to formulas, causing those who suffer from formularitis to congratulate themselves that they are all through with that group of cases and do not have to worry about them any more. Everyone tries to get some general rules to go by and so avoid the need of thinking things out from the beginning each time. It is popular to have a formula telling what to do, when to do it and how.

This is not a special failing of engineers; it is a common human trait, today, yesterday, ever. By the use of formulas people expect to get the maximum results with the minimum of time, effort and, especially, of responsibility. If the formula is wrong, that's not their fault; if they misunderstand it, that's because it isn't clear anyway.

And as this is argued, devils gibber and angels weep. In real life the formulas do not work very well. There are lots of such rules in the wise saws of the people, in the epigrams dear to poor Richard, in the advice to Laertes of that dear old bore Polonius; the early twentieth century was plagued with them.

The formulas are applicable when they apply and are useful when they work; that's all. An engineer claims that he was promoted for telling the chief engineer that he (the chief) didn't know what he was doing, and another man claims that he established a record in his office by admitting that he (the man) didn't know what he was doing; neither method is recommended as a general formula.

In fact there is no general formula for success because you are you and the other fellow is an entirely different animal. What is success for one man is a rather trivial accomplishment for another. What seems success at six is not attainment at sixty; men of forty do not all wish they had been firemen, or policemen, even though many still cherish an occasional secret ambition to chuck their professions and be Daniel Boone.

If people know just what they want, they can probably get it. But they'll have to pay for it. They may have to sacrifice peace or comfort or happiness or honor or friends or liberty. The trouble is that most people don't want to pay the price; they want to have their cake and eat it too. They think the fellow next door had his cake and ate it. It can't be done; they must always pay. Formularitis, though extremely common and sometimes epidemic, is rarely incurable in engineers; vigorous mental exercises in the fresh air of natural phenomena is recommended.

Translatitis is imported. It consists in exaggerating the value, importance and credibility of facts because they came from a considerable distance and were translated into English with some effort. Of course, it is true that facts bearing on any work should be at hand from the laboratories and literatures of all countries; it is not always possible, but it is desirable. However, quite unconsciously as a rule, many tend to measure the value of information by the distance from which it came and the effort devoted to its translation, as if engineering bore any similarity to postage stamps or tropical orchids. A leading engineer once tried to find the basis for an important rule which was at variance with usual practice. He was able only to learn that one member of the committee which formulated it had seen a statement in a certain foreign book that tests supported that rule, but the committee could not find the tests.

Perhaps the case just cited was complicated by experimentalitis. Experiments are very helpful, but a few or even many experiments may tell little. There is no field of study that requires more careful training or a keener intellect than the devising and interpreting of experiments. The shortest road to a fish fry of engineering facts is not promiscuous, indiscriminate experimentation—a process of dynamiting the pond of knowledge. Many tests give few facts and unless well devised they give none that anyone can be sure of. It is not good to eat fish all messed up with mud and driftwood. Except for the work of a few men of peculiar genius in the interpretation of test data, the least valuable part of any report of tests is the conclusions. To use those data safely, each man should draw his own conclusions. A more general tendency to do so would discourage the amateurish idea that this is an easy way to acquire knowledge and would further discourage the very objectionable custom of merely stating that tests indicate thus-and-so without explaining how the tests were made or how they showed what they are supposed to have shown. Students are prone to refer to tests when they can neither describe them nor even imagine tests to prove the alleged fact. What they generally mean is that they have seen or heard it stated that tests prove it and that they know nothing else about the matter.

Engineers get their information from several distinct sources: from their own experience in observing the action of natural forces or human customs and from records of observations by others; from mathematical analysis or models corresponding to such analyses; from experiments on the properties of materials or on structures or machines; from hunches and common sense; from weighing, interpreting, correlating and using such information. Experience is a guide which may be miscellaneous, fragmentary, unsatisfactory, often secondhand, frequently inaccurate, but no engineer will discount its tremendous importance as evidence.

All nature is trying to tell something of how its forces act. The best information, the most valuable material, comes directly from nature. Men may try to duplicate her phenomena in a laboratory but we never exactly reproduce the true natural problem, never fully ferret out her secrets. The greatest engineers are undoubtedly those who best learn to speak the language of nature.

Mathematical analysis in every field is dependent on assumptions. The structural engineer must accept certain conditions concerning elastic or plastic action. He must consider what goes on at working loads and also what conditions exist prior to failure. Engineers should put down some figures here, perhaps write some equations, but always remember that they are getting only some of the evidence in the case. This procedure may be frankly approximate and descriptive, as it usually is, or it may seek greater precision by the use of advanced mathematics.

The statistical method is recognized by scientists and engineers as a tool which may be dangerous if used carelessly.

Unfortunately its dangers are often forgotten and its misuse
has led to many errors. Those who have gone astray, however, have done so not by drifting into Mark Twain's
group of climatic liars, but by failing to remember how
pointedly true in engineering is Josh Billings's advice that
"It's better not to know so much than to know so many things
that ain't so."

There is an unfortunate tendency to burden engineers,
through books, with endless techniques and procedures of
mathematical analysis. Few students know that at best books
can furnish only a perishable net of large mesh through
which they may begin to strain their information and that
every fiber of that net must be rewoven from man's own
thinking and that many new strands must be added if it is
to be permanent and reliable in holding the selected data of
years of engineering practice. Books present the sets of tools;
it is the task of the analytical engineer to select those tools
which can be used most advantageously.

Of course, there can be no discrepancy between correct
theory and good practice. However, theories are not entirely
correct, because they are based on assumptions which limit
their application; and a theory which will not work is a
wrong theory. Yet all engineering is dependent on theory,
for it is only by theory that the profession can correlate
experience or interpret experiments; burning down a house
to roast a pig is too expensive. All theory is limited in application, but it cannot be dispensed with by the relation of

cause and effect, by experience or experiment, by neglecting it or even by common sense. Common sense is only the application of theories which have grown and been formulated unconsciously as a result of experience. But those who assume that the first thing to be done with an engineering problem is to begin industriously computing areas, moments and stresses will appear as absurd as did the little jurors in *Alice in Wonderland* who began busily to add up all the dates in the evidence and reduce the sum to pounds, shillings and pence.

Laboratory experiments may give valuable evidence. Engineers cannot take structures into the laboratories, but they can get evidence in laboratories with regard to the action of the structures. The multiplicity of factors involved is a source of great difficulty; many specimens of many types must be made and tested in many ways. The genius of experimentation must devise experiments that do not involve, in their interpretation, a theory more doubtful than that which the experiment was intended to investigate. There is a bad tendency in this field of study to drift into statements such as "Tests show that this is true." The more cautious engineers state and mean that these tests show that sometimes this is true, or even more cautiously that the results of these tests are not opposed to this conclusion. Exactly the same thing may be said of analytical procedures or of the experiments now popular with models. Engineers know that analyses, whether mathematical or by models, experiments and experience are all merely evidence bearing upon their problem, to be judiciously weighed in drawing conclusions.

All these sources of evidence provide needed information. The ability to correlate this knowledge and season it with dependable common sense is the rarest, the most valuable and the most difficult skill for an engineer to acquire. Sense of proportion, judgment of relative value cannot be learned from books, though books may guide in its attainment; teachers cannot guarantee it, though they may hasten its development; it will not automatically come with any length or variety of schooling or experience. Men must give it to themselves and if they do acquire it, it will bear the mark of their own individuality.

The idea that common sense is a gift of the gods is overdone. Some men will never have common sense in engineering problems; but it can be developed to some extent by those who work hard and hopefully and then repeatedly look over the whole field in which they have worked. They must try to see the hills and valleys, try to appreciate what parts are important and what parts are less important, try to be synthetic as well as analytic, to give due attention to probability, to develop some sense of relative importance. To these men will come in time what seems an intuitive faculty of getting the answer. Common sense provides a rapid qualitative approach to problems. In the hands of many it is a powerful source of evidence. It is true that many think they have it who lack it. The fact that it is dangerous does not make it either necessary or desirable to abandon it or to neglect it.

In studying an unfamiliar structural type, engineers may find all stresses under all loading conditions. Then they need a knowledge of the properties of the materials in the structure and no one may be quite prepared to say what these properties are. It is never conclusive and rarely easy to tell a manufacturer of steel or of aluminum what properties structural engineers require in their metal. They will not find all the properties of the material because they must define before they find and imagine before they define, which presupposes that rare animal—a good imaginer. And after finding these stresses and these properties, the engineer must study seriously the probable type of failure and combination of loads causing it.

Much of the best work of engineers is the result of hunches, vague analogies to other cases with which they have worked. It is undoubtedly true that good results come from hard work, but it is also strangely true that they often come from hard work done at some other time on some other problem. Hard work has a surprising way of paying unexpected dividends through later inspirations. However, one must clearly realize that hunches, because they are vague and formless and unreasoned, are dangerous. An analogy is not a reason—*comparison n'est pas de raison*—nor does similarity constitute identity. The idea suggested may prove true, or it may be nonsense; and yet the persistent hunch of a trained thinker should not be treated lightly. One does in time develop what has been called, with needless erudition, a "power of unconscious ratiocination."

So there is evidence from several sources. Rarely does this evidence all completely agree. None of the sources is in itself more dependable, more scientific, more satisfactory than any other one. All have at times given tremendous aid; all have at times grossly misled. It is necessary to do here what humanity has always done in its practical relations. The evidence must be assembled and importance given to the part that judgment indicates is most dependable.

Thoughtful engineers weigh the findings presented to them through all or any one of these sources with a full appreciation of the effect their personal prejudices might have on conclusions drawn from the evidence. Any man over forty has acquired so large a junk pile of prejudices, preconceptions, biases, convictions, notions, loves and hates that it is very hard for him to tell why he thinks what he thinks. It's tremendously hard at any age to be honest; it's hard for men when they are young because, though they have few prejudices, they also have few data, and it's harder later because they then have acquired bias as fast as or faster than they have gotten facts.

Ideas which men think they have created and of which they are so proud, on art or on science or with regard to literary forms or styles, are often merely depraved and impov-

erished hang-overs—hand-me-downs—from worked over
Grecian notions in the European renaissance or from Francis
Bacon and the Cartesian revolution or from Scott and the
romanticists or Addison or Smollett.

In Europe the river problem has been largely that of
navigation, not of floods, so their literature has been influenced, at first avowedly and later unconsciously, by a
desire to make the streams floatable. The Mississippi River
Commission at first had to justify its existence as a Federal
body on the grounds that it sought to improve the waterway to
the Gulf.

It is not that writers and investigators of flood control
have been morally dishonest but that they have often been
intellectually disingenuous, borrowing from this school of
thought or that group of interests, ideas and theories to the
support of which they molded their facts. This is so common
a practice that it often may be expected and when a man has
expressed one opinion some people jump to the conclusion
that they can predict all his opinions—and sometimes they
can.

There is, of course, a certain advantage in being prejudiced. It gives men something to start from on the voyage
or something to tie to if a storm comes and they want to
stay in port. Some people are so devoted to the ideal of forming unprejudiced opinions that, when they start to study a
subject, they carefully avoid reading anything in that field
or discussing the subject with others. The result is that their

conclusions may be just as much prejudiced, but the prejudices are all their own. However, what may be lauded as an
unprejudiced frame of mind, breadth of view, intellectual
liberalism, is often the most arrant twaddle—anemic intellectual sponginess.

On the other hand, while freedom from prejudice and pre-
conception are practically impossible, it is very important to
recognize and identify one's own personal prejudices, especially in engineering work. Engineers deal invariably with
both human ways and natural forces; their work is both a
product of and a foundation for the civilization and culture
of the race. But civilization and culture are not built in a
day. Some conclusions and opinions in engineering have
been inherited from a professor who studied under some
other professor who got his ideas from a German scholar—
and so the house that Jack built. On the other hand there
was a distinguished engineer who designed an approach up
a steep hill in an Eastern city—a technically excellent solution of a difficult problem—in such a way that it marred the
view of an old and loved church of which the whole town
was proud—a conspicuous neglect of prejudice.

All engineers have passed through recurrent periods of
conflict between what may be called the "practical" and the
"theoretical" approaches to engineering problems. Some
who think themselves practical show little sympathy for
analytical investigations. Their attitude is that they know by
divine intuition and experience how to do their job and they

do not consider that many details of this job are subject to a completely rational analysis. As opposed to these are those popularly conceived to be typical college professors, who think it possible to rationalize every procedure and to reduce it to rigid rules.

It may be agreed that in the field of structural engineering—perhaps some will even agree in the field of government—there is no need to adopt exclusively either the point of view of rugged individualism or that of planned economy. A judicious combination of the two is necessary. The individualist is, however, both by temperament and training, somewhat unfitted for planning, and the theorist is quite commonly unfitted for bold and imaginative excursions into new types of construction.

Many articles purporting to be new appear in the field of analysis. Sometimes such articles are useful; often they are harmful. Very much needed are methods of thinking in the analytical field that utilize the language and preserve the concepts familiar to constructors and men who have a natural and intuitive gift for imagining structural action. The burden here seems to lie on the theorists rather than on the practical men; they must meet the practical men more than halfway. In the field of civil engineering the designers and builders are the men on the firing line.

Analytical procedures in mechanics should be so simple and flexible that they may give quickly either a quantitative or a qualitative method of thinking. They should draw a picture of a structure in action. Great builders for thousands of years have necessarily formed in their minds some such pictures. The probability is that if someone tried to explain some of the "new" modern concepts to Michelangelo or to Peter of Colechurch or Galileo they would easily grasp the procedure. As a matter of fact, it would not be surprising if they replied that they knew the method all along.

For formal analysis, methods may be used that are not primarily methods of thinking at all. These are often very formalistic, like a sausage grinder. If certain numerical data are fed into one end of the analysis and a crank is turned, a lot of little sausages—moments, reactions, stresses, movements—come inevitably out of the other end of the machine. It works quite smoothly; in fact it works with deceiving smoothness. Because the sausages seem uniform and regular, it is often assumed that the meat cannot be spoiled.

New Lamps for Old

NOVELTY OR LIGHT

"Alike fantastic."

ENGINEERING has passed from the rugged individualism of, say, 1850 through a fairly judicious
combination of rugged individualism and planning of perhaps 1900 into an era in which much emphasis is put on
analysis. There are three quite distinct approaches to problems in engineering—analytical, experimental, synthetic.
None of these can progress independent of the others and
none of them should become subservient to the others. Engineers may become entirely too practical for the good of the
profession; analysts may become too theoretical, too ab-
struse. It is even more dangerous if the analysts become too
practical and the engineers too theoretical.

Consideration of prejudices is neither more nor less important than dependability of facts. Truth does not always
come labeled as such and quite frequently some so-called
scientific facts—all dressed up in dress suits and top hats
—are not genuine. Fallacies—illusions of great truths or
seductive novelties—may be compared to leading ladies and
chorus girls. Engineers must remember that it may be all
right to flirt with the chorus ponies provided you don't marry
them. Some may be very nice girls and some grow up to be fat

and sensible, but the main thing is not to marry them or at any rate not to marry too many of them or, anyway, not to marry too many of them too hastily. In other words, careers must not be irrevocably tied up, early or late, with new and pretty but untried theories, however interesting. It is the young men who must maintain extreme caution since most older engineers are too intellectually bald to start more flirtations.

In the engineer's world, the world of practical affairs, life is very real and very earnest and the goal is clearly defined. The function of engineering is to produce human wealth, which really means human comfort.

But to identify the facts, truths, laws which must precede this production of wealth is difficult and often disappointing. Much distraction comes from fallacious theories advertised by this school of thought or that group of thinkers.

Few systems of thought are free from fallacies, but theories based on fallacies are not necessarily wrong. In 1890 we knew that eating melons in the patch on a hot day was likely to cause malaria, which is very true unless a mosquito net is worn, but the form of the dogma involved a fallacy. In fact it is probable that most thinking either involves fallacies—defects of logic—or is closely associated with them. Someone has said that the whole theory of structural design is built up by attributing impossible properties to non-existent materials.

Some fallacies are like sisters and aunts, familiar mem-

bers of the family, and intimate association with their faults
serves only to further endear their virtues. Others are of the
chorus type, too pretty, too perfectly fascinating by their
novelty. They distract attention from their shallowness by a
lavish display of irrelevant extremities.

Many erudite fallacies are distortions of the views of some
great thinker, from whom lesser disciples borrow opera
glasses but fail to focus them for their own eyes. These disciples miss the great vision, the great purpose, and leave
to the world a detailed record of futility in seven volumes.
And those seven volumes pass into the hands of a number of
specialists, each of whom produces seven other volumes and
lays out a jigsaw puzzle that will never again fit together.

The great truths of engineering are simple; they can be
simply stated and simply applied. This is a very different
thing from saying that anyone has yet stated them simply or
showed how to apply them with ease. An endlessly complex
description or explanation of an engineering fact indicates
complications in the brain of the propounder rather than the
complexity of nature. Whatever cannot be stated in plain
English is half-baked, though no man may yet be able to fin-
ish the baking and half-baked is better than no bread. But
still what is half-baked is prolific of indigestion.

The field of structural engineering, for example, has recurrent periods of growing complexity, a piling of Pelions
of theory on Ossas of experiment; partial differentials pursuing herds of test data, fineness moduli and colloidal ratios

shriek gibberingly in the din of equations and diagrams and strain gage records. And out of it usually comes sanity and simplicity and better structures and materials,—some of the chorus have danced their last and some retire for a season.

The period of medieval scholasticism stretches from the ninth to the fourteenth century. It was a strange period, when wise men solemnly discussed the logical attributes of omniscience. And toward its end came John Duns Scotus, leader of the Scotist school of Franciscan scholars, Thomas Aquinas of the Dominicans, and Roger Bacon, forerunner of modern science.

Scotus was a fellow of Merton College, Oxford, doctor and dean of theology of the University of Paris; his defense of the doctrine of the Immaculate Conception finally led the University of Paris to require all candidates for the doctorate to forswear Thomist and Dominican errors. The subtle doctor lightly brushed aside the immature irrelevancies of his subordinates.

Roger Bacon also graduated from Oxford and Paris and joined the Franciscans. He listed the four causes of error as follows: authority, custom, the opinion of the unskilled mass of men, concealment of real ignorance with pretense of knowledge. Of these he says that the last is the most dangerous and the cause of all the others. He was forbidden to teach at Oxford.

It is accepted that Roger Bacon's thesis has a certain appositeness today and he is hailed as the precursor of modern scientific curiosity. And the name of Duns Scotus, the great dean of theology at the University of Paris, has been retained in our language, for a stupid fool is now called a "dunce."

But this was six centuries ago and today people are much wiser. Or are they?

Error always remains, part and parcel of the intellectual life. As Mr. Roget would phrase it, people have errors and fallacies, misconceptions, misapprehensions, misunderstandings, misinterpretations, misjudgments, heresies, misstatements, mistakes, faults, blunders; errata, delusions, illusions, hallucinations, absurdities, imbecilities, stupidities, puerilities, senilities, fatuities and nonsense. All of us make them, live by them and thrive on them.

The great intellectual tragedy is not in the chorus of fallacies nor with the beaux who flirt at the stage door. The stage-door Johnnies usually suffer from a damnable malady the name of which is youth, but nearly everyone who lives long enough gets over it.

The tragedy, the real tragedy, is with the Johnnies who marry one of the chorus. Young men should go to the intellectual music hall if they will and look 'em over, even sit in the front seat through one performance of "The Fallacies of 1952." However, they should be careful to pin their faith on something more enduring than paint and powder and periwigs, forms or formulas or fancies. When they feel sure of the soundness of some new theory, new method, new material, new type of structure, new machine, they should take their new idea on more than one buggy ride before they see a justice of the peace.

There is an old adage that says any fool can ask a question that the wisest cannot answer. A more important statement is that only the very wise can ask questions in such a way that any fool can answer them. If the questions are good questions the answer can probably be found and if they are poor questions no one can answer them.

The question of many children, "What does God look like?" is a poor question; it implies that He is not God. But there have been many pictures of gods, many images. If the little children try to draw pictures of God as they see Him they may revise their question and ask, "What is God?" —quite another question.

. For many years now a particular question has been asked in various ways: "What is rigidity?" "Why is it desirable for structures to be rigid?" "Is it always desirable for structures to be rigid?" "What is the proper measure of rigidity?" "Is the measure always the same?"

From time immemorial men have sought in their structures some property which they may call "rigidness" or "rigidity." Structural types are often said to have been selected on the basis of their relative rigidity. And yet after many years no one really knows what the word means.

Some engineers profess to know—they think they know, they think they think they know—that there is such a property; they do believe the cat they're looking for really exists though they're not quite sure what room he is in. Perhaps he is several cats, or it may be one cat living nine lives in different places. Yet the term and the synonyms of it are in such general use that we do know something about it; this cat is not entirely black.

A few rays of light may come from the synonyms. There are many words describing types of motion: wiggle, wobble, shiver, stagger, reel, roll, pitch, toss, gyrate, skip, hop, sway, shake, vibrate, twitch, switch, twist, bend, curl, jerk, squirm, wriggle, writhe, leap, bound, jump, swing, oscillate, wave, whirl, swirl, eddy, swish, tremble, waver, totter, quake, quiver.

People are obviously very conscious of types of movement, and what some people think all the time may not be significant and what everybody thinks sometimes may be in error but what all the people think all the time is important.

Of course engineers must look in all of the rooms because they're not quite sure whether they're talking about displacement, velocity, acceleration, or change of acceleration or all of them at once. They're not quite sure whether they're talking psychology—animal reaction to movement—or structural integrity and durability—the effect on a structure of movement. Probably they should look into the properties of materials as affected by shock or by repetition, and so they leave bunches of catnip around in the Materials Testing Laboratory.

But this is a consideration of questions, not rigidity. Until men ask the right question in the right way they'll not get far in studying rigidity, and when the questions are asked correctly the answer will probably be simple.

Pictures are the necessary supplement to questions. Students should be encouraged more to draw pictures of what they are talking about. They should draw pictures of deformed structures, pictures of structural failure, pictures of stress distribution. To try to draw them raises, or should raise, hundreds of questions. If men can't draw them they don't know what they are talking about and the degree of detail shows the amount of familiarity with the subject. To try to draw a picture, as the little children did, will frequently answer or invalidate a question.

Now there are many different types of pictures—photographs, cartoons, conventionalized diagrams. And there are many ways of drawing them—line diagrams, word pictures, mathematical descriptions, sketches. It is usually well and often necessary for an engineer to draw them in several ways. No one can take a photograph of stress distribution but there are ways of drawing conventional pictures representing it. Much education and thought goes into such pictures.

The sum and substance of this is that men's technical knowledge can be sized up better from the questions asked than from the answers given, and answers can be evaluated best by the pictures that accompany them. The first demand that the profession makes is for pictures. But never before in any field of technical study was there greater need for men who can ask the right questions.

Textbooks rarely ask important questions. Few professors do. The texts and the professors are too busy telling what they know to emphasize what they don't know. But the latter is often more important—to know the limitations of knowledge and to ask questions, simple questions, that may open up holes through which light can filter into our dark rooms. Only when we try to draw the pictures do we ask these questions, and we must ask them because we find that so much of the landscape to be painted is as yet hidden from view.

Lights in the Ivory Tower

FAITH AND HOPE—PERHAPS SOME CHARITY

"Engineering is the art of directing the great sources of power in nature for the use and convenience of man."

THE INSTITUTE of Civil Engineers of Great Britain was organized over a century ago. At that time "civil engineer" meant any engineer not formally engaged in military work. Thomas Tredgold, a successful practitioner and well-known writer on engineering topics, was asked to write a definition of the term; his statement was adopted and is today printed on all publications of the Institute. "The art of directing the great sources of power in nature for the use and convenience of man"; nothing better has been written for the purpose. It is still a good definition for all branches of the profession.

"For the use and convenience of man." This is as important a part of Tredgold's definition as any. Note the nice distinction between use and convenience; they are not always identical. Engineering does not try to tell men what they should want or why they want it. Rather it recognizes a want and tries to meet it. Hence engineers, perhaps more than other men, are interested in man, are interested in what men want and how men live and how men react to their environment.

Usefulness and convenience are relative terms. Obsolescence results from changing degrees of use and convenience. To take an oft-repeated example, the automobile in America has made many highways neither useful nor convenient.

A problem of continuing importance is the use and control of water. This assembles economic factors, charts of flow, prediction of rainfall and flood, structural problems of the design of dams, hydraulic problems of the control of water in canals or in turbines, and all of these reach far back into details of investigation in pure and applied science.

Engineering is devoted to the "use and convenience of man." As man's needs and desires have changed so has the art of engineering progressed, and consequently the historical development of the United States and of the world illustrates well the advance of engineering.

The history of engineering in America, and to a considerable degree the history of America itself, may be traced in terms of successive obstacles imposed by nature on the westward march of the people. First settlers on the East coast developed the harbor facilities and, in a rather unsuccessful way, the routes of transportation by land and water near the coast; by 1800 they met the great Allegheny escarpment. This barrier stretches from the Canadian border almost to the Gulf, a hundred miles or so from the Atlantic. There are only a few breaks, one through the Mohawk-Hudson depression, another in the low gaps between the headwaters of the Susquehanna and the Allegheny. In Virginia the headwaters of the James lead to those of the New River. Sherman's army followed another gap. A most dramatic story is that of the fight of Charleston to reach the west country. Boston was cut off by the Berkshires, but New York found a travel route of low grades at its door, pioneer rail lines crawled through the gaps as Philadelphia reached through Pennsylvania at about the same time that Baltimore completed the Baltimore and Ohio into Wheeling. The Chesapeake and Ohio reached up the valley of the James toward the Ohio.

Development of transportation follows the succession of available facilities and modes of travel. There were coastal canals, local canals, through canals such as the Erie. In time the relatively inefficient canal gave way to the railway, which brought with it problems of track and equipment and terminals. Each of these subjects has since been elaborated in detail by specialists. Track, for example, has received much technical investigation involving study of flexure, of the strength of soils and of rail fastenings, of ties, of the underlying roadbed itself. Track has become one of the critical problems of railway systems.

Westward-marching America finally reached the great valley with its far-flung system of floatable rivers. The heyday of the river steamboat was short; the railway, in the hands of brilliant engineers of that period, proved a more efficient servant of the people.

Then there was the rail crossing of the Mississippi. James B. Eads built the great structure that is still a model of grace and a marvel of technical work. The early dreams of

Mississippi River crossings, which seemed so impossible of achievement, were finally realized. After Eads came bridges at Memphis and Thebes, Cairo, Cape Girardeau, Vicksburg and finally at New Orleans.

In time there were crossings of the Missouri, tunnels and grades through and over the Rockies and the Sierras. Railways followed approximately the lines of the Oregon and Santa Fe Trails.

The history of America is here, the Erie Canal or the building of Brooklyn Bridge, the opening of the Baltimore and Ohio, the direct railway connection of St. Louis with the East, the building of the Central Pacific, the Soo Canal, the conquest of floods on the Miami, the development of the Tennessee Valley, the Woolworth tower, the Empire State Building, the George Washington or Transbay Bridges or that at Golden Gate.

Toward the end of the last century this country had a great volume of work to be done, but had few standards and analytical procedures or experimental data with which to work. Young men went abroad in the 70's and 80's to secure these. Most of them went to Germany and brought back elaborated products of German scholarship and America had to digest this technical formality. To use a homely phrase, America had to chew all of these fish and spit out the bones, and frequently the bones stuck in her throat. The German mind has a tendency to elaboration, often to complexity and recurrently a lack of discrimination in interpreting evidence. Structural literature in America at the beginning of the century was cluttered with a great deal of undigested technique, and much of it was bad. Some of it has been thrown away, and more of it should be discarded. However, it did serve its purpose because at that time it removed to some extent, though not by any means sufficiently, certain limitations on evidence.

Men sent to Europe in the 1920's to study laboratory methods in hydraulics absorbed more rapidly and with more discrimination. The greatest hydraulics laboratories today without exception are in America, thanks to the wholesome and mellowing influence of Yankee common sense.

In some fields of engineering that stage has now passed. Certainly in the future it will not be necessary to turn to Europe for technical thought; Americans will catch their own fish. Perhaps they will become more expert in spitting out the bones than they have been in the past.

Civilization still lives on a frontier, but the type of frontier changes. Anyone who travels over America realizes how much there remains to be done; it will take men of quantitative sense, trained to think precisely where precision is justified and trained also to know the limitations of precision.

Henceforth American destiny will be thought out at home. Britain and the Continent will still be a source of many thoughts and dreams, but the American people from now on must develop concepts of culture, methods of thought and philosophies of service along lines traditional in America.

The factory system, the beginnings of modern transportation, systematic studies of water supply and sanitation appear early in our history. America did not impose modern methods on a medieval society but grew through and with its engineers. It is important to recognize this in interpreting the "American way of life" and contrasting it with the problems of sections where the benefits of good engineering are not familiar. It may appear that today history is more in need of philosophic concepts from engineering than is engineering in need of historical perspective.

More development of natural resources will take place in the next twenty than has taken place in the past fifty years. To originate and plan this work it is important that somewhere there be men with a very clear understanding of the physical limitations placed by nature on the activities of men, limitations as to what can be provided for the "use and convenience of man."

In an obvious and dramatic sense our frontiers passed before the beginning of the century, but some people seem to have been rather slow in realizing it. The frontier of engineering, however, never passes; the problems are as insistent today as they were a hundred years ago. Harbors, rivers, sea beaches are to be cleaned up for better living, highway systems must be revised, better control of floods and of pollution of streams is urgent. There is much—very much—still to be done. The use and convenience change but the art itself does not appreciably change. This art reaches into every aspect of human relations. Robert Louis Stevenson, son and grandson of engineers, says of it:

"My grandfather was above all things a projector of works in the face of nature, and a modifier of nature itself. A road to be made, a tower to be built, a harbour to be constructed, a river to be trained and guided in its channel—these were the problems with which his mind was continually occupied, and for these and similar ends he travelled the world for more than a half of century, like an artist, note book in hand.

"What the engineer most properly deals with is that which can be measured, weighed, and numbered . . . not only entries in note books, to be hurriedly consulted; in the actor's phrase, he must be stale in them; in a word of my grandfather's, they must be 'fixed in the mind like the ten fingers and ten toes.'

"These are the certainties of the engineer; so far he finds a solid footing and clear views. But the province of formulas and constants is restricted. . . . With the civil engineer, more properly so called (if anything can be proper with this awkward coinage), the obligation starts with the beginning. He is always the practical man. The rains, the winds and the waves, the complexity and the fitfulness of nature, are always before him. He has to deal with the unpredictable, with those forces (in Smeaton's phrase) that 'are subject to no calculation'; and still he must predict, still calculate them, at his peril. His work is not yet in being, and he must foresee its influence: how it shall deflect the tide, exaggerate the waves, dam back the rainwater, or attract the thunderbolt . . . he must not only consider that which is, but that which may be.

"It is plain there is here but a restricted use for formulas. In this sort of practice, the engineer has need of some transcendental sense. . . . The rules must be everywhere indeed: but they must everywhere be modified by this transcendental coefficient, everywhere bent to the impression of the trained eye and the feelings of the engineer."

Most of the economic, industrial and cultural life of America lies within a zone between parallels Thirty-five and Forty-five north. The Fortieth Parallel passes through Philadelphia, Wheeling, Columbus, Springfield, Illinois, follows the Kansas-Nebraska border, passes through Denver and reaches the Pacific at Cape Mendocino just north of San Francisco. This zone extends on the East coast from Eastport to Hatteras (to quote the fascinating phraseology of the weather bureau), from Memphis to Minneapolis, from Salem to Santa Barbara. The Fortieth Parallel seems to be particularly congenial to humanity; in Europe this congenial zone is pushed northward along the west coast by the warmth of the Gulf Stream and hence runs diagonally northwest and southeast from Scandinavia to the Levant. If the map of the United States were laid down on that of Europe, it would extend from Belfast to Baghdad and from Madrid to Moscow. This would cover practically all of that part of

Europe represented by the civilization of the Near East and of western Europe. In Asia the Fortieth Parallel passes through Peiping, and not far from Tokyo.

Along this belt—U.S. Route 40—Americans have met most of their engineering problems—transportation, sanitation, industrial development, problems of "forge and farm and mine and bench." To solve these problems engineers experiment and analyze, chart the past to predict the future, plot data geological, climatological, meteorological, hydrological, pathological, study theories of physics and chemistry. Somehow they put them together for the use and convenience of man. On the whole the job has been well done; it is, at least, a marvel to other nations.

Engineering deals with man in his natural environment, with machines as substitutes for man and with the power to drive those machines, with materials and their methods of manufacture. From this rather obvious classification of men, machines and materials, there arises, in the academic world and in professional classifications, a breaking down into all sorts of specialties. The study of man in his natural environment would probably best fit into what is now commonly called civil engineering; machinery and power are represented by mechanical and electrical engineering; the development of new materials by the fields of metallurgy and chemical engineering. In some universities there are departments of ceramic engineering, agricultural engineering, aeronautical engineering and so on indefinitely.

In general the problems of civil engineers are given to them by God Almighty. They are problems in nature. On the other hand, mechanical and electrical work has problems which man, to a certain extent, has created for himself. This difference is in some ways fundamental to the type of problems to be studied, the method of studying it and the control over the result after the problem has been studied.

In primitive society the family needs shelter, water, some paths for getting food to the cabin and some way of disposing of its refuse. In a slightly more complex state of society it may wish to use the water to turn a mill wheel. Civil engineering builds the shelter, gets the water, opens the path. Wind and wave, flood and fire, earthquake and landslide, mud and rock and the eternal pull of gravity; man's need to sleep and work in safety and comfort, store and transport harvests, travel quickly and without danger by sea and by land, drink pure water and live in sanitary surroundings—these supply the problems of civil engineers.

Engineers study both human needs and natural phenomena; they must predict how much it will rain, how much of this rain they can store and where, but also they must know how much water people need and how many people will need it. These two fields of study give essential unity to the profession, for all engineers, whatever their specialty, must know both human ways and natural forces. Their work is to control and tame these forces. The civil engineers, in turn, are ever dependent on mechanical and electrical engineers to supply them with machines for accomplishing these ends and on chemical and metallurgical engineers to produce the necessary materials of construction.

Engineering is, and has always been, coexistent with civilization. Palaces and walls of Nineveh or Babylon, pyramids of dead Pharaohs, wharves of Mediterranean merchants, harbors of the Hanseatic League, Roman aqueducts and roads, bridges of Tiber or Thames, Rhine or Hudson, Hoover Dam or Grand Coulee, the Cloaca Maxima of Rome, Panama Canal, Galveston sea wall or Mississippi jetties, steel mill and weave shed, warehouse and workshop, the imagination races with the kaleidoscopic picture—each tells a story of the kind of men who lived and how they lived and why they lived and what they knew and how they thought. All the monuments that man has reared to prince or potentate, to Zeus, Jupiter or Jehovah, to commerce, industry or pleasure, were dreams and plans of the men who built them and record the history of the race, the progress of civilization, the foundations of today and tomorrow.

The legions march again, the Ecclesiastic procession enters, the holiday crowds roll by in their automobiles. Most men think of this as an immediate accomplishment, but it is rather the result of a long, gradual development which began back at the dawn of history.

Poor engineering entails failure and misfortune, inconvenience, suffering and death. Overestimate of available power of a stream or of the yield for water supply, imper-

fect sanitary provisions, poor location or construction of
transportation routes, unsafe bridges and buildings, power
plants without a market, railways without traffic; eventually
each of us pay the bill for these errors in money, convenience
or health. Errors of judgment will occur; we live in a world
of misunderstandings and misinterpretations, misjudgments
and mistakes. For just this reason competent engineers, on
guard against errors, go back to test their conclusions by
simple truths; for the great principles of engineering are
always simple, can be simply stated and simply applied,
though in some fields no one may have achieved this
simplicity. The simple perception and application of these
truths characterize those whose work has been of distinction.

As the evening ferries leave lower Manhattan, the details of the great buildings fade into the darkness and the
splendor of a fairy city shines out, the graceful towers of
the Woolworth and Singer and the broad fronts of the
Equitable or Whitehall seem to float out in glory, windows of
the upper stories traced by the lights of late workers. Below
are steel columns and girders and grillages and concrete
caissons to the schist a hundred feet underground. Within
the island run tubes, tunnels, sewers, conduits, subways;
all planned to the inch, functioning to the minute. Here lie
the entrails and there towers the head of civilization.

It may be objected that such a civilization stands on feet
of clay. Perhaps, but if modern sanitation or transportation

lead to materialism, let this civilization make the most of them; read Defoe's *Journal of the Plague Year* or travel narratives of the eighteenth century. Artist and poet have sought this vision—the genius of mankind brooding over nature and making chaos fertile, invoking the power of the Almighty to shield life and goods and home against torrent and tempest, famine and pestilence; both artist and poet often fail to portray vividly the pageant of human progress because they misconceive the anatomy of the forms they would delineate.

We may argue forever as to the relative "breadth" of professional activities, of studies of men's souls or minds or bodies or customs or language. It is not very important whether engineering is called a craft, a profession or an art; under any name this study of man's needs and of God's gifts that they may be brought together is broad enough for a lifetime.